APPLIED NUMERICAL METHODS:

AN INTRODUCTION

AUTHORS

Dr. N. Pratyusha

Mrs. P. Amaleswari

Dr. B. Krishna Veni

Dr. P. Bharath Kumar

FIRST EDITION

2024

PREFACE

Numerical methods form the foundation of modern scientific computation, enabling us to approximate solutions to complex mathematical problems that often cannot be solved analytically. From engineering and physics to biology, economics, and beyond, the application of numerical methods has grown significantly, fueled by the increasing power and accessibility of computational tools. This book, Numerical Methods, is designed to provide students and professionals with a solid foundation in both the theory and practical application of these essential techniques. The primary aim of this text is to bridge the gap between mathematical concepts and real-world application. It offers a comprehensive introduction to core numerical methods, focusing on topics such as root-finding algorithms, interpolation, numerical differentiation and integration, solving systems of linear equations, and ordinary differential equations. Each chapter is structured to emphasize both understanding the mathematical principles and developing hands-on problem-solving skills, with numerous examples and exercises designed to build intuition and confidence.

Sincerely,

Dr. N. Pratyusha

Mrs. P. Amaleswari

Dr. B. Krishna Veni

Dr. P. Bharath Kumar

ACKNOWLEDGMENTS

I am grateful to all those who contributed to the development of this book, which covers fundamental and advanced topics in numerical methods. This work would not have been possible without the valuable insights, encouragement, and expertise from many individuals.

I would like to extend my appreciation to my mentors and colleagues who provided guidance on topics such as **Algebraic and Transcendental Equations**, **System of Linear Equations**, and **Eigenvalue Problems**. Their input greatly enhanced the sections on methods like the **Iteration**, **Secant**, **Newton-Raphson**, **Regula-Falsi**, and **Bisection methods**, as well as the **Gauss-Seidel iteration**, **LU-Decomposition**, and **Thomas Algorithm**.

My sincere thanks also go to those who reviewed and contributed to sections on **Interpolation** and its various approaches, including **Newton's forward and backward interpolation formulas**, **Lagrange's interpolation formula**, **Spline interpolation**, and **Divided Difference methods**. Their insights helped make these topics more accessible and comprehensive for readers.

For the chapters on **Numerical Differentiation and Integration**, I am indebted to colleagues who shared their knowledge on **Numerical Differentiation**, **Quadrature formulas**, **Romberg's method**, and **Gaussian integration**, all of which strengthened the book's mathematical rigor.

I am also thankful to those who supported the sections on the **Numerical Solution of Ordinary Differential Equations**, including the **Runge-Kutta**, **Predictor-Corrector**, and **Adams-Bashforth-Moulton methods**. Their expertise helped ensure clarity and accuracy in these complex topics.

Finally, I wish to thank my family and friends for their patience, encouragement, and unwavering support throughout the process. I am equally appreciative of the editorial team for their diligent work, which helped bring this book to its final form.

Thank you to everyone who has contributed to this endeavor.

With gratitude,

[Dr. N. Pratyusha

Mrs. P. Amaleswari

Dr. B. Krishna Veni

Dr. P. Bharath Kumar]

Contents

UNIT-I

Algebraic and Transcendental Equations

1.1 Introduction and scope

Using mathematical modeling, most of the problems in Engineering and Sciences can be formulated in terms of systems of linear or non-linear equations, ordinary or partial differential equations, or integral equations. In the majority of the cases, the analytical methods used to solve the above-said equations may be difficult lengthy, or non-existent. For these problems, numerical methods are available. Numerical methods have always been useful and their role in engineering design and scientific research is of fundamental importance because they can give a solution for a problem when ordinary analytical methods fail. In recent years, the widespread use of computers in engineering and scientific research has made the study of numerical methods as important as the study of calculus for engineers and scientists. But they always provide approximate solutions.

In scientific and engineering work, a frequently occurring problem is to find the roots of the equation of the form $f(x) = 0$.

If $f(x)$ is quadratic cubic or bi-quadratic, then we have familiar methods for finding the roots, but always provide $f(x)$ may not be quadratic it may be algebraic or transcendental.

Definition An equation $f(x) = 0$ where $f(x) = a_0 x^n + a_1 x^{n-1} + a_2 x^{n-2} + ... + a_n$, $a_0, a_1, a_2, ... a_n$ are real numbers, $a_0 \neq 0$, is called an nth-degree algebraic equation, i.e., $f(x)$ is purely a polynomial.

If $f(x)$ contains some other functions namely trigonometric, logarithmic, exponential, etc., then the equation $f(x) = 0$ is called a transcendental equation. $f(x) = 0$

Root-A real number α is called the real root of the equation $f(x) = 0$ if and only if $f(\alpha) = 0$

Geometrically, the real root of is the value of x at which the graph of $f(x)$ meets the x-axis.

We can find the roots of the algebraic or transcendental equation by using numerical methods.

For example, $x^4 - 4x^2 + 5 = 0, 4x^2 - 5x + 7 = 0; 2x^3 - 5x^2 + 7x + 5 = 0$ are algebraic equations.

An equation that contains polynomials, trigonometric functions, logarithmic functions, exponential functions, etc., is called a Transcendental equation. For example,

$$\text{Tan}\, x - e^x = 0$$

$$\sin x - xe^{2x} = 0$$

$$x\, e^x = \cos x$$

are transcendental equations.

Finding the roots or zeros of an equation of the form $f(x) = 0$ is an important problem in science and engineering. We assume that f (x) is continuous in the required interval. A root of an equation $f(x) = 0$ is the value of x, say x = α for which $f(\alpha) = 0$. Geometrically, a root of an equation $f(x) = 0$ is the value of x at which the graph of the equation $y = f(x)$ intersects the x–axis (see Fig. 1.1)

2

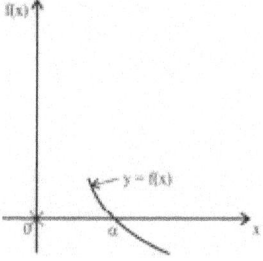

Fig. 1.1 Geometrical Interpretation of a root of $f(x) = 0$

A number α is a simple root of $f(x) = 0$; if $f(\alpha) = 0$ and $f'(\alpha) \neq 0$. Then, we can write $f(x)$ as,

$$f(x) = (x - \alpha)g(x), g(\alpha) \neq 0$$

A number α is a multiple root of multiplicity m of $f(x) = 0$

if $f(\alpha) = f'(\alpha) = \cdots .. = f^{(m-1)}(\alpha) = 0$ and $f^m(\alpha) = 0$.

Then, $f(x)$ can be written as,

$$f(x) = (x - \alpha)^m g(x), g(\alpha) \neq 0$$

A polynomial equation of degree n will have exactly n roots, real or complex, simple or multiple. A transcendental equation may have one root or no root or an infinite number of roots depending on the form of $f(x)$.

The methods of finding the roots of $f(x) = 0$ are classified as,

1. Direct Methods   2. Numerical Methods.

Direct methods give the exact values of all the roots in a finite number of steps. Numerical methods are based on the idea of successive approximations. In these methods, we start with one or two initial approximations to the root and obtain a sequence of approximations $x_0, x_1, \ldots \ldots \ldots x_k$ which in the limit as $k \to \infty$ converge to the exact root $x = a$.

3

There are no direct methods for solving higher degree algebraic equations or transcendental equations. Such equations can be solved by Numerical methods. In these methods, we first find an interval in which the root lies. If *a and b* are two numbers such that *f (a) and f (b)* have opposite signs, then a root of *f (x) =* *0* lies in between *a and b*. We take *a or b* or any valve in between *a or b* as first approximation x_1 . This is further improved by numerical methods. Here we a discuss few important Numerical methods to find a root of *f (x) = 0*.

1.2 ITERATION METHOD

Iteration is the process in which we perform the same procedure again and again. In the iterative method, we can solve the problem by calculating the successive approximations of the solution with an initial guess. Iteration methods are useful for problems involving a large number of variables.

Let $f(x) = 0$ be the given equations, it can be expressed as $x = \phi(x)$, let x_0 be an approximate value of the desired root α .

We calculate $x_1 = \phi(x_0)$

The successive approximations are approximately equal.

$x_{n+1} = \phi(x_n)$ is called the iterative formula.

Here one question arises.

Does the sequence of approximation $x_0, x_1, x_2, \ldots x_n$ always converge to the same number α ?

i.e., if the sequence of $\{x_n\}$ converges to α , then we can say that the iteration process is convergent. This method is convergent if $|\phi(x)| < 1$, $\forall, x \in I$, where I is the interval which contains a root of the equation.

Example.1.1 Find the root of the equation $\cos x = 3x - 1$ correct to 4 decimal places using an iterative method.

Solution Given $\cos x = 3x - 1$

$$Let\ f(x) = \cos x - 3x + 1$$

$$f(0) = 2$$
$$f(0.5) = 0.3375$$
$$f(0.6) = 0.025 > 0$$
$$f(0.7) = -0.335 < 0$$

Root lies between 0.6 to 0.7.

$$\cos x = 3x - 1$$

$$3x = 1 + \cos x$$

$$x = \frac{1 + \cos x}{3} = \phi(x) \Rightarrow \phi'(x) = \frac{-\sin x}{3}$$

$$|\phi'(x)| = \left|\frac{-\sin x}{3}\right| < 1, \forall x \in (0.6, 0.7)$$

Iterative method is applicable and the iterative formula is

$$x_{n+1} = \phi(x_n) \Rightarrow x_{n+1} = \frac{1 + \cos x_n}{3}.$$

Put n=0, $x_0 = 0.6$

$$x_1 = \frac{1 + \cos x_0}{3} = 0.6084$$

$$x_2 = \frac{1 + \cos x_1}{3} = 0.6068$$

$$x_3 = \frac{1 + \cos x_2}{3} = 0.6071$$

$$x_4 = \frac{1 + \cos x_3}{3} = 0.6071$$

$$\Rightarrow Root = 0.6071$$

Example.1.2 Find the root of the equation $x^3 + x^2 - 1 = 0$ using iterative method.

Solution

$$Let\ f(x) = x^3 + x^2 - 1$$
$$f(0.6) = -0.424 < 0$$
$$f(0.7) = -0.167 < 0$$
$$f(0.8) = 0.152 > 0$$

Root lies between 0.7 and 0.8.

$$x_1 = \left(1 - x_0^2\right)^{1/3} = 0.7113,\ x_{11} = \left(1 - x_{10}^2\right)^{1/3} = 0.7113$$

$$x_2 = \left(1 - x_1^2\right)^{1/3} = 0.7905,\ x_{12} = \left(1 - x_{11}^2\right)^{1/3} = 0.7654$$

$$x_3 = \left(1 - x_2^2\right)^{1/3} = 0.7211,\ x_{13} = \left(1 - x_{12}^2\right)^{1/3} = 0.7454$$

$$x_4 = \left(1 - x_3^2\right)^{1/3} = 0.7829,\ x_{14} = \left(1 - x_{13}^2\right)^{1/3} = 0.7631$$

$$x_5 = \left(1 - x_4^2\right)^{1/3} = 0.7287,\ x_{15} = \left(1 - x_{14}^2\right)^{1/3} = 0.7475$$

$$x^3 + x^2 - 1 = 0$$
$$x^3 = 1 - x^2$$
$$x = \left(1 - x^2\right)^{1/3} = \phi(x)$$
$$\phi'(x) = \frac{1}{3}\left(1 - x^2\right)^{\frac{1}{3} - 1} . 2x = \frac{-2x}{3\left(1 - x^2\right)^{2/3}}$$

$$\left|\phi'(x)\right| = \left|\frac{-2x}{3\left(1 - x^2\right)^{2/3}}\right| < 1, \forall, x \in (0.7, 0.8)$$

Implies iterative method is applicable and the iterative formula is
$$x_{n+1} = \phi(x_n)$$

$$x_{n+1} = \left(1 - x_n^2\right)^{1/3}$$

Put n=0, $x_0 = 0.8$.

$$x_6 = \left(1 - x_5^2\right)^{1/3} = 0.7769, x_{16} = \left(1 - x_{15}^2\right)^{1/3} = 0.7613$$
$$x_7 = \left(1 - x_6^2\right)^{1/3} = 0.7346, x_{17} = \left(1 - x_{16}^2\right)^{1/3} = 0.7491$$
$$x_8 = \left(1 - x_7^2\right)^{1/3} = 0.7721, x_{18} = \left(1 - x_{17}^2\right)^{1/3} = 0.7599$$
$$x_9 = \left(1 - x_8^2\right)^{1/3} = 0.7391, x_{19} = \left(1 - x_{18}^2\right)^{1/3} = 0.7504$$
$$x_{10} = \left(1 - x_9^2\right)^{1/3} = 0.7684$$

$\therefore$ Root of the equation=0.7504

Example.1.3 Find the cube root of 15, correct to 4 significant figures by iterative method.

Solution Iterative formula,

$$x_{n+1} = \phi(x_n) \Rightarrow x_{n+1} = \frac{15 - x_n^3 + 20x_n}{20}$$

$$n = 0, x_0 = 2.4$$

$$x_1 = \frac{15 - x_0^3 + 20x_0}{20} = 2.4655$$

$$x_2 = \frac{15 - x_1^3 + 20x_1}{20} = 2.4661$$

$$x_3 = \frac{15 - x_2^3 + 20x_2}{20} = 2.4662$$

Root=2.4662.

Example.1.4 Find a real root of $2x - \log_{10} x = 7$ by the iteration method.

Solution The given equation can be written as,

$$x = \frac{1}{2}(\log_{10} x + 7)$$

Let $x_0 = 3.8$

$$x_1 = \frac{1}{2}(\log_{10} 3.8 + 7) = 3.79$$

$$x_2 = \frac{1}{2}(\log_{10} 3.79 + 7) = 3.7893$$

$$x_3 = \frac{1}{2}(\log_{10} 3.7893 + 7) = 3.7893.$$

Therefore, x = 3.7893 is a root of the given equation which is correct to four significant digits.

Example.1.5 Find the root of the equation $2x = \cos x + 3$ correct to three decimal places using Iteration method.

Solution Given equation can be written as $x = \frac{(\cos x + 3)}{2}$

$$|\phi'(x)| = \left|\frac{\sin x}{2}\right| < 1$$

Hence iteration method can be applied

$$\text{Let } x_0 = \frac{\pi}{2}$$

$$\therefore x_1 = \frac{1}{2}\left(\cos\frac{\pi}{2} + 3\right) = 1.5$$

$$\therefore x_2 = \frac{1}{2}(\cos 1.5 + 3) = 1.535$$

Similarly, $x_3 = 1.518,$

$$x_4 = 1.526,$$

$$x_5 = 1.522,$$

$$x_6 = 1.524,$$

$$x_7 = 1.523,$$

$$x_8 = 1.524.$$

The required root is x = 1.524

Example.1.6 Using the method of iteration find a positive root between 0 and 1 of the equation $x\,e^x = 1$

Solution The given equation can be written as $x = e^{-x}$

Here $|\phi'(x)| < 1$ for x< 1.

9

We can use iterative method

Let $x_0 = 1$

$$x_1 = e^{-1} = \frac{1}{e} = 0.3678794.$$

$$x_2 = e^{0.3678794} = 0.6922006.$$

$$x_3 = e^{0.6922006} = 0.5004735$$

Proceeding like this, we get the required root as $x = 0.5671$.

1.3. SECANT METHOD

A potential problem in implementing the Newton-Raphson method is the evaluation of the derivative. Although this is not inconvenient for polynomials and many other functions, there are certain functions whose derivatives may be extremely difficult or inconvenient to evaluate. For these cases, the derivative can be approximated by a backward finite divided difference.

$$f'(x_i) \cong \frac{f(x_{i-1}) - f(x_i)}{x_{i-1} - x_i}$$

$$x_{i+1} = x_i - \frac{f(x_i)(x_{i-1} - x_i)}{f(x_{i-1}) - f(x_i)}$$

This approximation can be substituted to yield the following iterative equation.

Example.1.7 Use the secant method to estimate the root of $f(x) = e^{-x} - x$. Start with initial estimates of $x_{-1} = 0$ and $x_0 = 1.0$

Solution Recall that the true root is 0.56714329.

10

First iteration

$$x_{-1} = 0$$
$$x_0 = 1$$
$$f(x_{-1}) = 1.00000$$
$$f(x_0) = -0.63212$$
$$x_1 = 1 - \frac{0.63212}{1 + 0.63212} = 0.61270$$
$$\varepsilon_t = 8.0$$

Second iteration

$$x_0 = 1$$
$$f(x_0) = -0.63212$$
$$x_1 = 0.61270$$
$$f(x_1) = -0.07081$$

Here both estimates are now on the same side of the root.

$$x_2 = 0.61270 - \frac{-0.07081(1 - 0.61270)}{-0.63212 - (-0.07081)} = 0.56384$$

$$\varepsilon_t = 0.58$$

Third iteration

$$x_1 = 0.61270$$
$$f(x_1) = -0.07081$$
$$x_2 = 0.56384$$
$$f(x_2) = 0.00518$$

$$x_3 = 0.56384 - \frac{0.00518(0.61270 - 0.56384)}{-0.07081 - (-0.00518)}$$

$$= 0.56717$$

$$\varepsilon_t = 0.0048$$

1.4. NEWTON-RAPHSON METHOD

The Newton-Raphson method is a more advanced method in finding the root of the equation $f(x) = 0$. It is used to improve the result obtained by bisection or Regula-Falsi method.

Let $f(x) = 0$ be the given equation. Let x_0 be the approximate root of $f(x) = 0$. If x_1 is the exact root of the equation, then

$$f(x_1) = 0 \text{ and } f(x_1) = f(x_0 + (x_1 - x_0)) = 0,$$

By Taylor's series,

$$f(a+h) = f(a) + hf'(a) + \frac{h^2}{\angle 2} f''(a) + ...$$

$$f(x_1) = f(x_0) + (x_1 - x_0) f'(x_0) + \frac{(x_1 - x_0)^2}{\angle 2} f''(a) + ... = 0$$

Suppose $x_1 - x_0$ is very small, then the higher powers can be neglected.

$$\therefore f(x_0) + (x_1 - x_0) f'(x_0) = 0$$

$$\Rightarrow x_1 - x_0 = \frac{-f(x_0)}{f'(x_0)}$$

$$Illy, x_2 = x_1 - \frac{f(x_1)}{f'(x_1)}$$

12

Generally,

$$x_{n+1} = x_n - \frac{f(x_n)}{f'(x_n)}$$

This is the Newton-Raphson formula.

Note. Newton-Raphson formula converges if the initial approximation. x_0 is chosen sufficiently close to the root x_1.

The root of the equation $f(x) = 0$ can be computed if the initial approximation x_0 satisfies the condition, $f(x_0).f''(x_0) > 0$.

1.4.1. Geometrical significance

Let the graph of $f(x) = 0$ is drawn.

A tangent is drawn at a point. $P(x_n, f(x_n))$ to the curve $y = f(x)$. This tangent meets the x-axis at $(x_{n+1}, 0)$.

Equation of the tangent at $P(x_n, f(x_n))$ is

$$y - y_n = \left(\frac{dy}{dx}\right)_{(x_n, y_n)} (x - x_n)$$
$$y - f(x_n) = f'(x_n)(x - x_n)$$

Since $(x_{n+1}, 0)$ lies on above equation, on substituting we get

$$x_{n+1} = x_n - \frac{f(x_n)}{f'(x_n)}$$

Example.1.8 Find the root of the equation 2x-5=3sinx by Newton-Raphson method correct to 3 decimal places.

Solution   Given $f(x) = 2x - 5 - 3\sin x$

$$f(2.5) = -1.7954$$
$$f(2.6) = -1.3465$$
$$f(2.7) = -0.8821$$
$$f(2.8) = -0.404 < 0$$
$$f(2.9) = 0.0822 > 0$$

The root lies between 2.8 and 2.9.

$$f'(x) = 2 - 3\cos x$$

Newton-Raphson method, $x_{n+1} = x_n - \dfrac{f(x_n)}{f'(x_n)}$

$$x_{n+1} = x_n - \frac{2x_n - 5 - 3\sin x_n}{2 - 3\cos x_n}$$
$$n = 0, x_0 = 2.9$$
$$x_1 = 2.8832$$
$$n = 1, x_1 = 2.8832$$
$$x_2 = 2.8832$$

Root=2.8832

Example.1.9 Obtain Newton-Raphson formula to find cube root of N and hence deduce the cube root of 12.

Solution   Let

14

$$x = N^{1/3} \Rightarrow x^3 = N$$
$$f(x) = x^3 - N \Rightarrow f'(x) = 3x^2$$

Newton-Raphson formula, $\quad x_{n+1} = x_n - \dfrac{f(x_n)}{f'(x_n)}$

$$x_{n+1} = x_n - \frac{x_n^3 - N}{3x_n^2}$$

To find the root of 12, put N=12

$$x_{n+1} = \frac{2x_n^3 + 12}{3x_n^2}$$

Put n=0, $x_0 = 2.2$ in above equation

$$x_1 = 2.2931$$

Put $n = 1, x_1 = 2.2931$

$$x_2 = 2.2894,$$

If n=2, then $x_3 = 2.2894$

Root =2.2894.

Example.1.10 Find a root of the equation $x^2 - 2x - 5 = 0$ by the Newton-Raphson method.

Solution Here $f(x) = x^3 - 2x - 5$.

$$\therefore f^1(x) = 3x^2 - 2$$

Newton – Raphson method formula is

$$x_{n+1} = x_n - \frac{f(x_n)}{f'(x_n)}$$

$$x_{n+1} = x_n - \frac{x_n^3 - 2x_n - 5}{3x_n^2 - 2},$$

$$n = 0, 1, 2, ...$$

Let $x_0 = 2$.

$$f(x_0) = f(2) = 2^3 - 2(2) - 5 = -1$$

$$f^1(x_0) = f^1(2) = 3(2)^2 - 2 = 10$$

Putting n=0 in the above equation, we get

$$x_1 = 2 - \left(\frac{-1}{10}\right) = 2.1$$

$$f(x_1) = f(2.1) = (2.1)^3 - 2(2.1) - 5 = 0.061$$

$$f^1(x_1) = f^1(2.1) = 3(2.1)^2 - 2 = 11.23$$

$$x_2 = 2.1 - \frac{0.061}{11.23} = 2.094568$$

Similarly, we can calculate $x_3, x_4 ......$

Example.1.11 Find a root of $x \sin x + \cos x = 0$, using Newton-Raphson method.

Solution $f(x) = x \sin x + \cos x = 0$.

$$f'(x) = \sin x + x \cos x - \sin x = x \cos x$$

The Newton – Raphson method formula is,

$$x_{n+1} = x_n - \frac{x_n \sin x_n + \cos x_n}{x_n \cos x_n}, \qquad n = 0, 1, 2, ...$$

Let $x_0 = \pi = 3.1416$

16

$$x_1 = 3.1416 - \frac{3.1416 \sin \pi + \cos \pi}{3.1416 \cos \pi} = 2.8233.$$

Similarly,

$$x_2 = 2.7986, \quad x_3 = 2.798 \text{ and } \quad x_4 = 2.7984$$

Example.1.12 Find the smallest positive root of $x - e^{-x} = 0$, using Newton - Raphson method.

Solution Let $f(x) = x - e^{-x}$

$$f^1(x) = 1 + e^{-x}$$

$$f(0) = -1 \quad \text{and} \quad f(1) = 0.63212.$$

The smallest positive root of $f(x) = 0$ lies in between 0 and 1.

$$\text{Let } x_0 = 1$$

The Newton – Raphson method formula is,

$$x_{n+1} = x_n - \frac{x_n - e^{-x_n}}{1 + e^{-x_n}}, \quad n = 0, 1, 2, \dots$$

$$f(0) = f(1) = 0.63212$$

$$f'(0) = f'(1) = 1.3679$$

$$x_1 = x_0 - \frac{x_0 - e^{-x_0}}{1 + e^{-x_0}} = 1 - \frac{0.63212}{1.3679} 0.5379.$$

$$f(0.5379) = -0.0461$$

$$f'(0.5379) = 1.584.$$

$$x_2 = 0.5379 + \frac{0.0461}{1.584} = 0.567$$

Similarly, $x_3 = 0.56714$

17

$x = 0.567$ we can be taken as the smallest positive root of $x - e^{-x} = 0$, correct to three decimal places.

Example.1.13 Using the Newton-Raphson method (a) Find the square root of a number (b) Find a reciprocal of a number.

Solution Let n be the number

$$x = \sqrt{n} \implies x^2 = n$$

If $f(x) = x^2 - n = 0$ then
the solution to $f(x) = x^2 - n = 0$ is $x = \sqrt{n}$.

$$f'(x) = 2x$$

By Newton Raphson's method

$$x_{i+1} = x_i - \frac{f(x_i)}{f'(x_i)} = x_i - \left(\frac{x_i^2 - n}{2x_i}\right)$$

$$x_{i+1} = \frac{1}{2}\left(x_i + \frac{x}{x_i}\right)$$

using the above formula the square root of any number $'n'$ can be found to required accuracy.

(a) To find the reciprocal of a number $'n'$

$$f(x) = \frac{1}{x} - n = 0$$

∴ Solution of above equation is $x = \frac{1}{n}$

$$f'(x) = -\frac{1}{x^2}$$

Now by Newton – Raphson method,

18

$$x_{i+1} = x_i - \left(\frac{f(x_i)}{f'(x_i)}\right)$$

$$x_{i+1} = x_i - \left(\frac{\frac{1}{x^2} - N}{-\frac{1}{x_i^2}}\right)$$

$$x_{i+1} = x_i(2 - x_i)$$

Using the above formula, the reciprocal of a number can be found to the required accuracy.

Example.1.14 Find the reciprocal of 18 using the Newton–Raphson method.

Solution The Newton-Raphson method

$$x_{i+1} = x_i(2 - x_i n)$$

Considering the initial approximate value of x as $x_0 = 0.055$nd given $n = 18$

$$x_1 = 0.055[2 - (0.055)(18)]$$

$$x_1 = 0.0555$$

$$x_2 = 0.0555[2 - 0.0555 \times 18]$$

$$x_2 = (0.0555)(1.001)$$

$$x_2 = 0.0555$$

Hence $x_1 = x_2 = 0.0555$

∴ The reciprocal of 18 is 0.0555

Example.1.15 Find a real root for $x \tan x + 1 = 0$ using the Newton–Raphson method.

Solution Given $f(x) = x \tan x + 1 = 0$

$$f'(x) = x \sec^2 x + \tan x$$

$$f(2) = 2 \tan 2 + 1 = -3.370079 < 0$$

$$f(3) = 2 \tan 3 + 1 = -0.572370 > 0$$

$$\therefore \text{The root lies between 2 and 3}$$

Take $\quad x_0 = \frac{2+3}{2} = 2.5 \ $ (average of 2 and 3)

By Newton-Raphson method

$$x_{i+1} = x_i - \left(\frac{f(x_i)}{f'(x_i)}\right)$$

$$x_1 = x_0 - \left(\frac{f(x_0)}{f'(x_0)}\right)$$

$$x_1 = 2.5 - \frac{(-0.86755)}{3.14808}$$

$$x_1 = 2.77558$$
$$x_2 = x_1 - \frac{f(x_i)}{f'(x_i)};$$

$$f(x_1) = -0.06383,$$

$$f'(x_1) = 2.80004$$

$$x_2 = 2.7558 - \frac{(-0.06383)}{2.80004}$$

$$x_2 = 2.798$$

$$f(x_2) = -0.001080, \quad f'(x_2) = 2.7983$$

$$x_3 = x_2 - \frac{f(x_2)}{f'(x_2)} = 2.798 - \frac{(-0.001080)}{2.7983}$$

$$x_3 = 2.798$$

$$x_2 = x_3$$

$\therefore$ The real root of $x \tan x + 1 = 0$ is 2.798

Example.1.16 Find a root of $e^x \sin x = 1$ using the Newton–Raphson method.

Solution   Given $f(x) = e^x \sin x - 1 = 0$

$$f'(x) = e^x \sec x + e^x \cos x$$

Take $x_1 = 0, x_2 = 1$

$$f(0) = f(x_1) = e^0 \sin 0 - 1 = -1 < 0$$

$$f(1) = f(x_2) = e^1 \sin(1) - 1 = \ \ 1.287 > 0$$

The root of the equation lies between 0 and 1

Using Newton-Raphson method

$$x_{i+1} = x_i - \frac{f(x_i)}{f'(x_i)}$$

Now consider $x_0 =$ average of 0 and 1

$$x_0 = \frac{1+0}{2} = 0.5$$

$$x_0 = 0.5$$

$$f(x_0) = e^{0.5} \sin(0.5) - 1$$

$$f'(x_0) = e^{0.5} \sin(0.5) + e^{0.5} \cos(0.5) = 2.2373$$

$$x_1 = x_0 - \frac{f(x_0)}{f'(x_0)} = 0.5 - \frac{(-0.20956)}{2.2373}$$

$$x_1 = 0.5936$$

$$f(x_1) = e^{0.5936} \sin(0.5936) - 1 = 0.0128$$

$$f'(x_1) = e^{0.5936} \sin(0.5936) + e^{0.5936} \cos(0.5936) = 2.5136$$

$$x_2 = x_1 - \frac{f(x_1)}{f'(x_1)} = 0.5936 - \frac{(0.0128)}{2.5136}$$

$$\therefore \quad x_2 = 0.58854$$

Similarly, $\quad x_3 = x_2 - \dfrac{f(x_1)}{f'(x_1)}$

$$f(x_2) = e^{0.58854} \sin(0.58854) - 1 = 0.0000181$$

$$f'(x_2) = e^{0.58854} \sin(0.58854) + e^{0.58854} \cos(0.58854)$$

$$f(x_2) = 2.4983$$

$$\therefore \quad x_3 = 0.58854 - \frac{0.0000181}{2.4983}$$

$$x_3 = 0.5885$$

$$\therefore \quad x_2 - x_3 = 0.5885$$

0.5885 is the root of the equation $e^x \sin x - 1 = 0$

Example.1.17 Find a real root of the equation

$xe^x - \cos x = 0$ using the Newton-Raphson method.

Solution Given $f(x) = xe^x - \cos x = 0$

$$f'(x) = xe^x + e^x + \sin x = (x + 1)e^x + \sin x$$

Take $\quad f(0) = 0 - \cos 0 = -1 < 0$

$$f(1) = e - \cos 1 = 2.1779 > 0$$

∴   The root lies between 0 and 1

Let $x_0 = \dfrac{0+1}{2} = 0.5$ (average of 0 and 1)

Newton-Raphson method

$$x_{i+1} = x_i - \frac{f(x_i)}{f'(x_i)}$$

$$x_{i+1} = x_0 - \frac{f(x_0)}{f'(x_0)} = 0.5 - \frac{(-0.053221)}{(1.715966)}$$

$$x_1 = 0.5310$$

$$f(x_1) = 0.040734, \quad f'(x_1) = 3.110063$$

$$x_2 = x_1 - \frac{f(x_1)}{f'(x_1)} = 0.5310 - \frac{0.040734}{3.11064}$$

$x_2 = 0.5179;$

$$f(x_2) = 0.0004339, \; f'(x_2) = 3.0428504$$

$$x_3 = 0.5179 - \frac{(0.0004339)}{3.0428504}$$

$$x_3 = 0.5177$$

∴ $f(x_3) = 0.000001106$

$$f(x_3) = 3.04214$$

$$x_4 = x_3 - \frac{f(x_3)}{f'(x_3)} = 0.5177 - \frac{0.000001106}{3.04212}$$

$$x_4 = 0.5177$$

∴ $x_3 = x_4 = 0.5177$

∴ The root of $xe^x - \cos x = 0$ is 0.5177

Note: A method is said to be order P or has the rate of convergence P, if P is the largest positive real number for which there exists a finite constant $c \neq 0$, such that

$$|\epsilon_{k+1}| \leq c$$

Where $\epsilon_K = x_K - \xi$ is the error in the Kth iterate. C is called Asymptotic Error constant and depends on derivative of $f(x)$ at x= ξ. It can be shown easily that the order of convergence of Newton–Raphson method is 2.

Note: (i) The approximation x_{n+1} given by (3) converges, provided that the initial approximation x_0 is chosen sufficiently close to the root of $f(x) = 0$.

(ii) Convergence of Newton-Raphson method: Newton-Raphson method is similar to the iteration method

$$\phi(x) = x - \frac{f(x)}{f(x)}$$

Differentiating the above equation w.r.t to 'x' and using condition for convergence of iteration method i.e.

$$|\phi'(x)| < 1,$$

we get

$$\left|1 - \frac{f'(x).f'(x) - f(x)f''(x)}{[f'(x)]^2}\right| < 1,$$

Simplifying we get condition for convergence of Newton-Raphson method is

$$|f(x).f''(x)| < [f(x)]^2.$$

1.5. REGULA-FALSI METHOD OR METHOD OF FALSE POSITION

This is the oldest method for finding the real root of an equation and closely resembles the bisection method.

Consider the equation $f(x) = 0$. Let x_0 and x_1 be two values of x such that $f(x_0)$ and $f(x_1)$ are of opposite signs. Since the graph of $y = f(x)$ crosses x-axis, the root must lie between x_0 and x_1.

The chord joining $A(x_0, f(x_0))$ and $B(x_1, f(x_1))$ meets x-axis. Equation of line AB,

$$y - f(x_0) = \frac{f(x_1) - f(x_0)}{x_1 - x_0}(x - x_0)$$

The point of intersection of the line with x-axis will be the first approximation of the root of $f(x) = 0$. Let it be $(x_2, 0)$ at this point y=0.

$$\Rightarrow 0 - f(x_0) = \frac{f(x_1) - f(x_0)}{x_1 - x_0}(x_2 - x_0)$$

$$\Rightarrow x_2 - x_0 = \frac{(x_1 - x_0)f(x_0)}{f(x_1) - f(x_0)}$$

If, $f(x_0), f(x_2)$ are of opposite signs, then root lies between x_0, x_2.

Put $x_1 = x_2$, then we get next approximation. If $f(x_1), f(x_2)$ are of opposite signs, then root lies between x_1, x_2.

Put $x_0 = x_2$, we get next approximation. We proceed in this way till the two successive approximations are approximately equal.

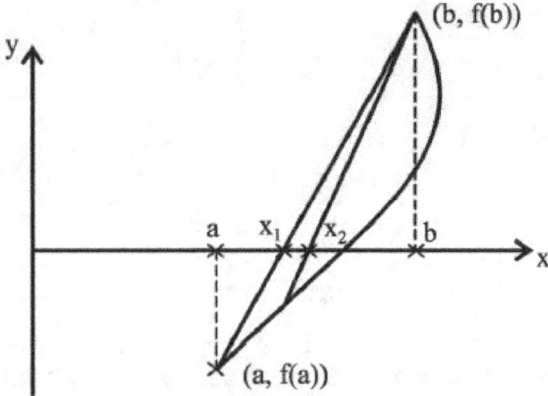

Fig.1.2. Regula falsi method

The formula is $x_{n+1} = x_{n-1} - \dfrac{(x_n - x_{n-1})}{f(x_n) - f(x_{n-1})} f(x_{n-1})$.

Example.1.18 Find the real root of the equation $x^3 - 9x + 1 = 0$ by Regula-Falsi method.

Solution Let

$$f(x) = x^3 - 9x + 1$$
$$f(2) = -9 < 0$$
$$f(3) = 1 > 0$$

Root lies between 2 and 3. Root is nearer to 3.

$$f(2.8) = -2.248.$$
$$f(2.9) = -0.711 < 0$$
$$f(3) = 1 > 0$$

Root lies between 2.9 and 3.

$$x_{n+1} = x_{n-1} - \frac{(x_n - x_{n-1})}{f(x_n) - f(x_{n-1})} f(x_{n-1})$$

Put $n = 1, x_0 = 2.9, x_1 = 3$

$$x_2 = x_0 - \frac{x_1 - x_0}{f(x_1) - f(x_0)} \cdot f(x_0)$$
$$= 2.9 - \frac{3 - 2.9}{f(3) - f(2.9)} f(2.9) = 2.9415$$
$$f(x_2) = f(2.9415) = -0.022 < 0$$

Root lies between 2.9415 and 3.

Put
$$n = 2, x_1 = 2.9415, x_2 = 3$$
$$x_3 = x_1 - \frac{x_2 - x_1}{f(x_2) - f(x_1)} \cdot f(x_1)$$
$$x_3 = 2.9415 - \frac{3 - 2.9415}{f(3) - f(2.9415)} f(2.9415)$$
$$x_3 = 2.9427$$
$$f(x_3) = -0.002 < 0$$

Root lies between 2.9427 and 3.

Put

$$n = 3, x_2 = 2.9427, x_3 = 3$$

$$x_4 = x_2 - \frac{x_3 - x_2}{f(x_3) - f(x_2)} \cdot f(x_2)$$

$$x_3 = 2.9427 - \frac{3 - 2.9427}{f(3) - f(2.9427)} f(2.9427)$$

$$x_3 = 2.9428$$

Root = 2.9428.

Example.1.19 Find the real root of the equation $x \log_{10}^x = 1.2$ correct to 4 decimal places.

Solution Given that

$$f(x) = x \log_{10}^x - 1.2$$
$$f(2) = -0.6 < 0$$
$$f(3) = 0.23 > 0$$

Root lies between 2 and 3.

$$x_{n+1} = x_{n-1} - \frac{(x_n - x_{n-1})}{f(x_n) - f(x_{n-1})} f(x_{n-1})$$

$$n = 1, x_0 = 2, x_1 = 3$$

$$x_2 = x_0 - \frac{x_1 - x_0}{f(x_1) - f(x_0)} \cdot f(x_0)$$

$$= 2 - \frac{3 - 2}{f(3) - f(2)} f(2) = 2.72$$

$$f(x_2) = f(2.72) = -0.04 < 0$$

Root lies between 2.72 and 3.

Put

$$n = 2, x_1 = 2.72, x_2 = 3$$

$$x_3 = x_1 - \frac{x_2 - x_1}{f(x_2) - f(x_1)} \cdot f(x_1)$$

$$x_3 = 2.72 - \frac{3 - 2.72}{f(3) - f(2.72)} f(2.72)$$

$$x_3 = 2.74$$

$$f(x_3) = -0.0006 < 0$$

Root lies between 2.74 and 3.

Put

$$n = 3, x_2 = 2.74, x_3 = 3$$

$$x_4 = x_2 - \frac{x_3 - x_2}{f(x_3) - f(x_2)} \cdot f(x_2)$$

$$x_3 = 2.74 - \frac{3 - 2.74}{f(3) - f(2.74)} f(2.74)$$

$$x_3 = 2.7407$$

Root = 2.7407.

Example.1.20 Find the root of the equation $2x - \log x = 7$ which lies between 3.5 and 4 by Regula–False method.

Solution Given that $f(x) = 2x - \log x_{10} = 7$

Take $x_0 = 3.5$, $x_1 = 4$.

Using Regula Falsi method

$$x_2 = x_0 - \frac{x_1 - x_0}{f(x_1) - f(x)} \cdot f(x_0)$$

$$x_2 = 3.5 - \frac{4 - 3.5}{(0.3979 + 0.5411)}(-0.5441)$$

$$x_2 = 3.7888$$

Now taking, $x_0 = 3.7888$ and $x_1 = 4$

$$x_3 = x_0 - \frac{x_1 - x_0}{f(x_1) - f(x_0)} \cdot f(x_0)$$

$$x_3 = 3.7888 - \frac{4 - 3.7888}{0.3988}(-0.0009)$$

$$x_3 = 3.7893$$
The required root is =3.789

Example.1.21 Find a real root of $xe^x = 3$ using Regula-Falsi method.

Solution Given $f(x) = xe^x - 3 = 0$

$$f(1) = e - 3 = -0.2817 < 0$$
$$f(2) = 2e^2 - 3 = 11.778 > 0$$
$\therefore$ One root lies between 1 and 2
Now taking, $x_0 = 1, x_1 = 2$
Using Regula – Falsi method
$$x_2 = x_0 - \frac{x_1 - x_0}{f(x_1) - f(x_0)} f(x_0)$$

$$x_2 = \frac{x_0 f(x_1) - x_1 f(x_0)}{f(x_1) - f(x_0)}$$
$$x_2 = \frac{1(11.778) - 2(-0.2817)}{11.778 + 0.2817}$$

$$x_2 = 1.329$$
$$f(x_2) = f(1.329) = 1.329e^{1.329} - 3 = 2.0199 > 0$$

$$f(1) = -0.2817 < 0$$

∴ The root lies between 1 and 1.329

Taking $x_0 = 1$ and $x_2 = 1.329$

$$\therefore \quad x_3 = \frac{x_0 f(x_2) - x_2 f(x_0)}{f(x_2) - f(x_0)}$$

$$= \frac{1(2.0199) + (1.329)(0.2817)}{(2.0199) + (0.2817)}$$

$$= \frac{2.3942}{2.3016} = 1.04$$

Now $f(x^3) = 1.04e^{1.04} - 3 = -0.05 < 0$

The root lies between x^2 and x^3

i.e., 1.04 and 1.329 $(\because f(x_2) > 0 \text{ and } f(x_3) < 0)$

$$\therefore \quad x_4 = \frac{x_2 f(x_3) - x_3 f(x_2)}{f(x_3) - f(x_2)}$$

$$= \frac{(1.04)(-0.05) - (1.329)(2.0199)}{(-0.05) - (2.0199)}$$

$x_4 = 1.08$ is the approximate root

Example.1.22 Find a real root of $e^x \sin x = 1$ using Regula – Falsi method.

Solution Given $f(x) = e^x \sin x = 1$

Consider $x_0 = 2$

$$f(x_0) = f(2) = e^2 \sin 2 - 1 = -0.7421 < 0$$

$$f(x_1) = f(3) = e^3 \sin 3 - 1 = 0.511 > 0$$

∴ The root lies between 2 and 3

Using Regula – Falsi method

$$x_2 = x_0 - \frac{x_0 f(x_1) - x_1 f(x_0)}{f(x_1) - f(x_0)}$$

$$x_2 = \frac{2(0.511) + 3(0.7421)}{0.511 + 0.7421}$$

$$x_2 = 2.93557.$$

$$f(x_2) = e^{2.93557} \sin(2.93557) - 1$$

$$f(x_2) = -0.35538 < 0$$

$\therefore$ Root lies between x_2 and x_1

i.e., lies between 2.93557 and 3

$$x_3 = \frac{x_2 f(x_1) - x_1 f(x_2)}{f(x_1) - f(x_2)}$$

$$= \frac{(2.93557)(0.511) - 3(-35538)}{0.511 + 0.35538}$$

$$x_3 = 2.96199.$$

$$f(x_3) = e^{2.90199} \sin(2.96199) - 1 = -0.000819 < 0$$

$\therefore$ Root lies between x_3 and x_1

$$x_4 = \frac{x_3 f(x_1) - x_1 f(x_3)}{f(x_1) - f(x_3)}$$

$$x_4 = \frac{2.9199(0.511) + 3(0.000819)}{0.511 + 0.000819}$$

$$= 2.9625898$$

$$f(x_4) = e^{2.9625898} \sin(2.9625898) - 1$$

$$= -0.0001898 < 0$$

$\therefore$ The root lies between x_4 and x_1

$$x_5 = \frac{x_4 f(x_1) - x_1 f(x_4)}{f(x_1) - f(x_4)}$$

$$= \frac{2.9625898(0.511) + 3(0.0001898)}{0.511 + (0.0001898)}$$

$$x_5 = 2.9626$$

We have $\qquad x_4 = 2.9625$

$$x_5 = 2.9626$$

$\therefore \qquad x_5 = x_4 = 2.962$

The root lies between 2 and 3 is 2.962

Example.1.23 Find a real root of $xe^x = 2$ using Regula – Falsi method.

Solution $f(x) = xe^x - 2 = 0$

$$f(0) = -2 < 0, \quad f(1) = i.e., -2 = (2.7183) - 2$$

$$f(1) = 0.7183 > 0$$

$\therefore$ The root lies between 0 and 1

Considering $x_0 = 0, x_1 = 1$

$$f(0) = f(x_0) = -2; \quad f(1) = f(x_1) = 0.7183$$

By Regula – Falsi method

$$x_2 = x_0 - \frac{x_0 f(x_1) - x_1 f(x_0)}{f(x_1) - f(x_0)}$$

$$x_2 = \frac{0(0.7183) - 1(-2)}{0.7183 - 2} = \frac{2}{2.7183}$$

$$x_2 = 0.73575$$

Now

$$f(x_2) = f(0.73575)$$

$$= 0.73575e^{0.73575} - 2$$

$$f(x_2) = -0.46445 < 0$$

And $f(x_1) = 0.7183 > 0$

$\therefore$ The root x_3 lies between x_1 and x_2

$$x_3 = \frac{x_2 f(x_1) - x_1 f(x_2)}{f(x_1) - f(x_2)}$$

$$x_3 = \frac{(0.73575)(0.7183)}{0.7183 + 0.4645}$$

$$x_3 = \frac{0.52848 + 0.46445}{1.18275}$$

$$x_3 = \frac{0.992939}{1.18275}$$

$$x_3 = 0.83951$$

$$f(x_3) = \frac{(0.83951)}{(0.83951)e^{-2}}$$

$$f(x_3) = (0.83951)e^{0.83951} - 2$$

$$f(x_3) = -0.056339 < 0$$

$\therefore$ One root lies between x_1 and x_3

$$x_4 = \frac{x_3 f(x_1) - x_1 f(x_3)}{f(x_1) - f(x_3)}$$

$$= \frac{(0.83951)(0.7183) - 1(-0.056339)}{0.7183 + 0.056339}$$

$$= \frac{0.65935}{0.774639} = 0.851171$$

$$f(x_4) = 0.851171e^{0.851171} - 2 = -0.006227 < 0$$

Now x_5 lies between x_1 and x_4

$$x_5 = \frac{x_4 f(x_1) - x_1 f(x_4)}{f(x_1) - f(x_4)}$$

$$x_5 = \frac{(0.851171)(0.7183) + (0.006227)}{0.7183 + 0.006227}$$

$$x_5 = \frac{0.617623}{0.724527} = 0.85245$$

$$f(x_5) = 0.85245e^{0.85245}e^{0.85245} - 2 = -0.0006756 < 0$$

$\therefore$ One root lies between x_1 and x_5,

(i.e., x_6 lies between x_1 and x_5)

Using Regula - Falsi method

$$x_6 = \frac{(0.85245)(0.7183) + (0.006756)}{0.7183 + 0.006756}$$

$$x_6 = 0.85260$$

Now $f(x_6) = -0.00006736 < 0$

$\therefore$ One root x_7 lies between x_1 and x_6

By Regula – Falsi method

$$x_7 = \frac{x_6 f(x_1) - x_1 f(x_6)}{f(x_1) - f(x_6)}$$

$$x_7 = \frac{(0.85260)(0.7183) + 0.0006736}{0.7183 + 0.0006736}$$

$$x_7 = 0.85260$$

From $x_6 = 0.85260$ and $x_7 = 0.85260$

$\therefore$ A real root of the given equation is 0.85260

Example.1.24 Determine the root of the equation $\cos x - x\,e^x = 0$ by the method of False position.

Solution Let $f(x) = \cos x - x\,e^x$

$f(0) = 1$ and $f(1) = -2.177979523$

$a = 0$ and $b = 1$. The root lies in between *0 and 1*

$$x_1 = \frac{0(-2.177979523) - 1(1)}{-2.177979523 - 1} = 0.3146653378$$

$$f(x_1) = f(0.314653378) = 0.51986.$$

The root lies in between 0.314653378 and 1.

$$x_2 = \frac{0.3146653378(-2.177979523) - 1(0.51986)}{-2.177979523 - 0.51986} = 0.44673$$

Proceeding like this, we get

$$x_3 = 0.49402,$$

$$x_4 = 0.50995,$$

$$x_5 = 0.51520,$$

$$x_6 = 0.51692,$$

Example.1.25 Determine the smallest positive root of $x - e^{-x} = 0$, correct of three significant figures using Regula False method.

Solution Let $f(x) = x - e^{-x}$

$f(0) = 0 - e^{-0} = -1$ and $f(1) = 1 - e^{-1} = 0.63212$,

The smallest positive root lies in between 0 and 1. Here a = 0 and b = 1

$$\therefore x_1 = \frac{0(0.63212) - 1(-1)}{0.63212 + 1} = 0.6127$$

$$f(0.6127) = 0.6127 - e^{-(0.6127)} = 0.0708$$

Hence, the next approximation lies in between 0 and 0.6127. Proceeding like this, we get

$$x_2 = 0.57219,$$

$$x_3 = 0.5677,$$

$$x_4 = 0.5672,$$

$$x_5 = 0.5671.$$

Hence, the smallest positive root, which is correct up to three decimal places, is,

$$x = 0.567$$

Example.1.26 Find a real root of the equation $f(x) = x^3 - 2x - 5 = 0$ by the method of False position.

Solution Let $f(x) = x^3 - 2x - 5$

$$f(2) = -1 \; and \; f(3) = 16$$

Hence the root lies in between *2* and *3*.

Take $a = 2$, $b = 3$.

$$x_1 = \frac{a\,f(b) - b\,f(a)}{f(b) - f(a)}$$

$$= \frac{2(16) - 3(-1)}{16 - (-1)} = \frac{35}{17} = 2.058823529.$$

$$f(x_1) = f(2.058823529) = -0.390799917 < 0.$$

Therefore, the root lies between 0.058823529 and 3. Again, using the formula, we get the second approximation as,

$$x_2 = \frac{2.058823529(16) - 3(-0.390799917)}{16 - (-0.390799917)} = 2.08126366$$

Proceeding like this, we get the next approximation as,

$$x_3 = 2.089639211,$$

$$x_4 = 2.092739575,$$

$$x_5 = 2.09388371,$$

$$x_6 = 2.094305452,$$

$$x_7 = 2.094460846,$$

1.6. Errors in numerical computations

In numerical computation, we work with a finite number of digits and carryout the computation in finitely many steps. Hence the numerical result so obtained is an approximate value of the unknown exact result.

Absolute error: If α is an absolute value of a quantity whose exact value is a, then the difference $E = \alpha - a$ is called the absolute error or simply error of α.

Relative error: It is denoted by E_r, it is defined as

$$E_r = \frac{E}{a} = \frac{\alpha - a}{a} \text{ provided } a \neq 0,$$

The percentage error, is defined by $E_p = 100 E_r$, $r = a - \alpha = -E$ is called the correction.

Thus $a = \alpha + r$

i.e., True value=approximate value +correction.

1.7. BISECTION METHOD

Consider the equation $f(x) = 0$, find a and b for which $f(a) < 0$ and $f(b) > 0$.

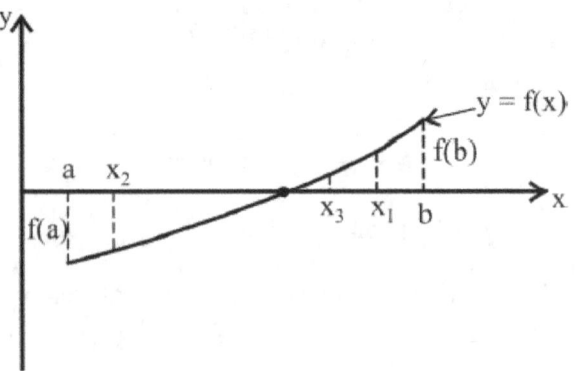

Fig.1.3.Bisection method

Find $x_1 = \dfrac{a+b}{2}$, calculate $f(x_1)$. If $f(x_1) = 0$ then x_1 becomes the root of the equation

$f(x) = 0$ otherwise

i.   If $f(x_1) < 0$, then root lies between x_1 and b.

Find $x_2 = \dfrac{x_1 + b}{2}$, calculate $f(x_2)$ and so on.

ii.   If $f(x_1) > 0$, then root lies between a and x_1.

Find $x_3 = \dfrac{a + x_1}{2}$, calculate $f(x_3)$ and so on.

We proceed in this way until the two successive approximation are approximately equal.

Example.1.27 Find the real root of $x^3 - x - 1 = 0$ using bisection method.

Solution Let $f(x) = x^3 - x - 1$

$$f(x) = x^3 - x - 1$$

$$f(1) = -1 < 0, f(2) = 5 > 0$$

Roots lies between 1 and 2.

$$x_1 = \frac{a+b}{2} = \frac{1+2}{2} = 1.5$$

$$f(1.5) = 0.875 > 0$$

Root lies between 1 and 1.5.

$$x_2 = \frac{1+1.5}{2} = 1.25,$$

$$f(x_2) = f(1.25) = -0.296 < 0$$

Root lies between 1.5 and 1.25

$$x_3 = \frac{1.5+1.25}{2} = 1.375,$$

$$f(x_3) = f(1.375) = 0.2246 > 0$$

Root lies between 1.25 and 1.375

$$x_4 = \frac{1.25 + 1.375}{2} = 1.3125,$$

$$f(x_4) = f(1.3125) = -0.015 < 0$$

Root lies between 1.3125 and 1.375

$$x_5 = \frac{1.3125 + 1.375}{2} = 1.3437,$$

$$f(x_5) = f(1.3437) = 0.0823 > 0$$

Root lies between 1.3437 and 1.3125

$$x_6 = \frac{1.3437 + 1.3125}{2} = 1.3281,$$

$$f(x_6) = f(1.3281) = 0.0144 > 0$$

Root lies between 1.3125 and 1.3281

$$x_7 = \frac{1.3125 + 1.3281}{2} = 1.3203,$$

$$f(x_7) = f(1.3203) = -0.018 < 0$$

Root lies between 1.3203 and 1.3281

$$x_8 = \frac{1.3203 + 1.3281}{2} = 1.3242,$$

$$f(x_7) = f(1.3242) = -0.0022 < 0$$

Therefore, root of the problem is 1.3242

Example.1.27 Find the root of the equation xsinx-1=0 lies in between x=1 and x=1.5.

Solution Let

$$f(x) = x \sin x - 1$$

$$f(1) = -0.1585 < 0$$

$$f(1.5) = 0.4962 > 0$$

Root lies between 1 and 1.5.

No. of iterations	a	b	$x_i = \dfrac{a+b}{2}$	$f(x_i)$
1	1	1.5	1.25	$0.1862 > 0$
2	1	1.25	1.125	$0.015 > 0$
3	1	1.125	1.0625	$-0.0718 < 0$
4	1.0625	1.125	1.09375	$-0.028 < 0$
5	1.09375	1.125	1.109375	$-0.0066 < 0$
6	1.109375	1.125	1.1175	$0.00408 > 0$
7	1.109375	1.1171	1.1132	$-0.0012 < 0$
8	1.1132	1.1171	1.11515	$0.0013 > 0$

Therefore, the root of the equation $=1.11515$

Example.1.28. Find a root of the equation $x^4 - x - 10 = 0$ using the Bisection method correct to 2 decimal places.

Solution Let $f(x) = x^4 - x - 10 = 0$ be the given equation.

We observe that $f(1) < 0$, then $f(2) > 0$.

So, one root lies between 1 and 2.

Let $x_0 = 1, x_1 = 2$;

Take $x_2 = \frac{x_0 + x_1}{2} = 1.5$; $f(1.5) < 0$;

∴ The root lies between 1.5 and 2

Let us take

$x_3 = \frac{1.5 + 2}{2} = 1.75$; we find that $f(1.75) < 0$,

∴ The root lies between 1.75 and 2

So, we take now

42

$$x_4 = \frac{1.75 + 1.875}{2} = 1.8125 = 1.81$$

can be taken as the root of the given equation.

Example.1.29 Find a real root of equation $x^3 - x - 11 = 0$ by Bisection method.

Solution   Given equation is $f(x) = x^3 - x - 11$

We observe that $f(2) = -5 < 0$ and $f(3) = 13 > 0$.

∴   A root of (1) lies between 2 and 3;

Take $x_0 = 2, \ x = 3$;

Let $x_2 = \frac{x_0 + x_1}{2} = \frac{2+3}{2} = 2.5$;

Since $f(2.5) > 0$, the root lies between 2 and 2.5

Taking $x_3 = \frac{2+2.5}{2} = 2.25$, we note that $f(2.25) < 0$;

The root can be taken as lying between 2.25 and 2.5.

∴ The root $= \frac{2.25 + 2.5}{2} = 2.375$

Example.1.30 Find a real root of $x^3 - 5x + 3 = 0$

using the Bisection method.

Solution   Let $f(x) = x^3 - 5x + 3 = 0$ be the equation given

Since $f(1) = -1 < 0$ and $f(2) = 1 > 0$, a real root lies between 1 and 2.

i.e., $x_0 = 1, x_1 = 2$;

Take $x_2 = \frac{1+2}{2} = 1.5; f(1.5) = -1.25 < 0$

43

∴   The root lies between 1.5 and 2;

$$\text{Take } x_3 = \frac{1.5+2}{2} = 1.75$$

$$f(1.75) = \left(\frac{7}{4}\right)^3 - 5\left(\frac{7}{4}\right) + 3 = -\text{ve};$$

∴ The root lies between 1.75 and 2

$$\text{Let} \quad x_4 = \frac{1.75+2}{2} = 1.875;$$

We find that $f(1.875) = (1.875)^3 - 5(1.875) + 3 > 0$

The root of the given equation lies between 1.75 and 1.875

$$\therefore \text{The root} = \frac{1.75+1.875}{2} = 1.813$$

Example.1.31 Find a real root of the equation $x^3 - 6x - 4 = 0$ by the Bisection method.

Solution   Here $f(x) = x^3 - 6x - 4 = 0$

$$x_0 = 2, x_1 = 3; \quad (\because f(2) < 0, f(3) > 0)$$

$$x_1 = 2.5; f(x_1) < 0;$$

$$\text{Take } x_3 = \frac{2.5+3}{2} = 2.75$$

$$f(2.75) > 0$$

$$\Rightarrow x_4 = \frac{2.5 + 2.75}{2} = 2.625$$

$$f(2.625) < 0$$

$$\Rightarrow \text{Root lies between 2.625 and 2.75}$$

Approximately the root will be $= \frac{2.5+2.75}{2} = 2.69$

Note: Solution of algebraic and transcendental equations .

The numerical methods to find the roots of $f(x) = 0$.

Bisection method: If a function $f(x)$ is continuous between a and b, $f(a)$ and $f(b)$ are of apposite sign then there exists at least one root between a and b. The approximate value of the root between them is $x_0 = \frac{a+b}{2}$

If $f(x_0) = 0$ then the x_0 is the correct root of $f(x) = 0$. If $f(x_0 0) \neq 0$, then the root either lies in between $x_0 \left(a, \frac{a+b}{2}\right)$ or $(\frac{a+b}{2}, b)$ depending on whether $f(x_0)$ is negative or positive. Again, bisection the interval and repeat same method until the accurate root is obtained.

Method of false position: (Regula false method): This is another method to find the root of $f(x) = 0$. In this method, we choose two points a and b such that $f(a), f(b)$ are of apposite signs. Hence the root lies in between these points $[a, f(a)], [b, f(b)]$ using equation of the chord joining these points and taking the point of intersection of the chord with the x-axis as an approximate root (using $y = 0$ on x– axis) is
$$x_1 = \frac{af(b) - bf(a)}{f(b) - f(a)}$$
Repeat the same process till the root is obtained to the desired accuracy.

Newton Raphson method: The successive approximate roots are given by
$$x_{n+1} = x_n - \frac{f(x_n)}{f'(x_n)}, n = 0, 1, 2 ---$$
provided that the initial approximate root x_0 is chosen sufficiently close to the root of $f(x) = 0$.

Exercise.1.1

1. Find a smallest positive root of $\tan x = x$ by Newton – Raphson. Method.

2. Find a root of $\sin x = 10(x - 1)$, using Iteration Method.

3. Find a real root of $\cot x = e^x$, using Iteration Method.

4. Find a root of $x^4 - x - 10 = 0$ by Newton – Raphson Method.

5. Find a real root $x - \cos x = 0$ by Newton – Raphson Method.

6. Find a root of $2x - 3 \sin x - 5 = 0$ by Newton – Raphson Method.

7. Using Method of False position, obtain a root of $x^3 + x^2 + x + 7 = 0$, correct to three decimal places.

8. Find the root of $x^3 - 2x^2 + 3x - 5 = 0$, which lies between 1 and 2, using Regula False method.

9. Compute the real root of $x \log x - 1.2 = 0$, by the Method of False position.

10. Find the root of the equation $\cos x - x e^x = 0$, correct to four decimal places by Method of False position

11. Using Iteration Method find a real root of the equation $x^3 - x^2 - 1 = 0$.

12. Find a real root of $\sin^2 x = x^2 - 1$, using iteration Method.

Unit-2
System of linear equations and eigenvalue problems

2.1. System of linear equations

In algebra, you learn to solve equations by first "simplifying" them using operations that do not alter the solution set. For example, to solve $2x = 8 - 2x$ we can add to both sides $2x$ and obtain $4x = 8$ and then multiply both sides by $1/4$ yielding $x = 2$. We can do similar operations on a linear system. There are three basic operations, called elementary operations, that can be performed:

1. Interchange two equations
2. Multiply an equation by a nonzero constant
3. Add a multiple of one equation to another.

These operations do not alter the solution set. The idea is to apply these operations iteratively to simplify the linear system to a point where one can easily write down the solution set. It is convenient to apply elementary operations on the augmented matrix [A b] representing the linear system. In this case, we call the operations elementary row operations, and the process of simplifying the linear system using these operations is called row reduction. The goal with row reducing is to transform the original linear system into one having a triangular structure and then perform back substitution to solve the system. This is best explained via an example

Example.2.1 Use back substitution on the augmented matrix

$$\begin{bmatrix} 1 & 0-2 & -4 \\ 0 & 1-1 & 0 \\ 0 & 01 & 1 \end{bmatrix}$$ to solve the associated linear system.

Solution Notice that the augmented matrix has a triangular structure. The third row corresponds to the equation $x_3 = 1$. The second row corresponds to the equation

47

$x_2 - x_3 = 0$ and $x_2 = x_3 = 1.$

The first row corresponds to the equation $x_1 - 2x_3 = -4$

$$x_1 = -4 + 2x_3 = -4 + 2 = -2.$$

Therefore, the solution is $\left(-2, 1, 1\right)$

Example.2.2 Solve the linear system using elementary row operations

$$-3x_1 + 2x_2 + 4x_3 = 12$$
$$x_1 - 2x_2 = -4$$
$$2x_1 - 3x_2 + 4x_3 = -3$$

Solution Our goal is to perform elementary row operations to obtain a triangular structure and then use back substitution to solve. The augmented matrix is

$$\begin{bmatrix} -3 & 2 & 4 & 12 \\ 1 & 0 & -2 & -4 \\ 2 & -3 & 4 & -3 \end{bmatrix}$$

Interchange row 1 and row2

$$\begin{bmatrix} -3 & 2 & 4 & 12 \\ 1 & 0 & -2 & -4 \\ 2 & -3 & 4 & -3 \end{bmatrix} \rightarrow \begin{bmatrix} 1 & 0 & -2 & -4 \\ -3 & 2 & 4 & 12 \\ 2 & -3 & 4 & -3 \end{bmatrix}$$

As you will see, this first operation will simplify the next step. Add 3*row1 to row2

$$\begin{bmatrix} 1 & 0 & -2 & -4 \\ -3 & 2 & 4 & 12 \\ 2 & -3 & 4 & -3 \end{bmatrix} \rightarrow \begin{bmatrix} 1 & 0 & -2 & -4 \\ 0 & 2 & -2 & 0 \\ 2 & -3 & 4 & -3 \end{bmatrix}$$

Add $-2R_1 + R_3$

$$\begin{bmatrix} 1 & 0 & -2 & -4 \\ 0 & 2 & -2 & 0 \\ 2 & -3 & 4 & -3 \end{bmatrix} \rightarrow \begin{bmatrix} 1 & 0 & -2 & -4 \\ 0 & 2 & -2 & 0 \\ 0 & -3 & 8 & 5 \end{bmatrix}$$

Add $R_2 + \dfrac{1}{2}$

$$\begin{bmatrix} 1 & 0 & -2 & -4 \\ 0 & 1 & -1 & 0 \\ 0 & -3 & 8 & 5 \end{bmatrix} \rightarrow \begin{bmatrix} 1 & 0 & -2 & -4 \\ 0 & 1 & -1 & 0 \\ 0 & -3 & 8 & 5 \end{bmatrix}$$

Add $3R_2 + R_3$

$$\begin{bmatrix} 1 & 0 & -2 & -4 \\ 0 & 1 & -1 & 0 \\ 0 & -3 & 8 & 5 \end{bmatrix} \rightarrow \begin{bmatrix} 1 & 0 & -2 & -4 \\ 0 & 1 & -1 & 0 \\ 0 & 0 & 5 & 5 \end{bmatrix}$$

Add $R_3 + \dfrac{1}{5}$

$$\begin{bmatrix} 1 & 0 & -2 & -4 \\ 0 & 1 & -1 & 0 \\ 0 & 0 & 5 & 5 \end{bmatrix} \rightarrow \begin{bmatrix} 1 & 0 & -2 & -4 \\ 0 & 1 & -1 & 0 \\ 0 & 0 & 1 & 1 \end{bmatrix}$$

We can continue row reducing but the row reduced augmented matrix is triangular So now

use back substitution to solve. The linear system associated with the row is reduced.

2.2. Gauss-Seidel iteration method

This is a modification of the Jacobi's iteration method. As before, we start with initial approximations, x_0, y_0, z_0 for x, y, *and z* respectively. Substituting $y = y_0, z = z_0$ in the first of the equations, we get, $x_1 = k_1$

Then putting $x = x_1, z = z_0$ in the second of the equations, we have $y_1 = k_2 - l_2 x_1 - m_2 z_0$

Next substituting in the third of the equation, we obtain $z_1 = k_3 - l_3 x_1 - m_3 y_1$ and so on. i.e., as soon as a new approximation for an unknown is found, it is immediately used in the next step. This process of iteration is continued till convergence to the desired degree of accuracy is obtained.

Observation.1. Since the most recent approximation of the unknowns are used while proceeding to the next step, the convergence in the Gauss-Seidal method is faster than in Jacobi's method.

Observation.2. The Gauss-Seidal method converges if in each equation, the absolute value of the largest coefficient is greater than the sum of the absolute values of the remaining coefficients.

Example 2.3   Solve Equations 2x+y=8,x+2y=1 using Gauss Seidel method

Solution Given that 2x+y=8,x+2y=1

From, the above equations

$$x_{k+1} = 12(8 - y_k)$$
$$y_{k+1} = 12(1 - x_k + 1)$$

Initial,gauss (x,y)=(0,0)

1*st* Approximation

$$x_1 = 4$$
$$y_1 = -1.5$$

2nd Approximation

$$x_2 = 1.75$$
$$y_2 = -1.875$$

3rd Approximation

$$x_3 = 4.9375$$
$$y_3 = -1.9688$$

4th Approximation

$$x_4 = 4.9844$$
$$y_4 = -1.9922$$

5th Approximation

$$x_5 = 4.9961$$
$$y_5 = -1.998$$

6th Approximation

$$x_6 = 4.9999$$
$$y_6 = -1.9995$$

7th Approximation

$$x_7 = 4.9998$$
$$y_7 = -1.9999$$

Solution of the Guass-Seidal method

$$x = 4.9998 \cong 5$$
$$y = -1.9999 \cong -2$$

Iterations are tabulated as below.

Iteration	x	y
1	4	-1.5
2	4.75	-1.875
3	4.9375	-1.9688
4	4.9844	-1.9922
5	4.9961	-1.998
6	4.999	-1.9995
7	4.9998	-1.9999

Example.2.4 Solve Equations 2x+5y=16,3x+y=11 using Gauss seidel method.Hence, we re-arrange the equations as follows, such that the elements in the   coeficient matrix are diagonally dominant.

Solution   Let   2x+5y=16   and   3x+y=11
The coefficient matrix of the given system is not diagonally dominant.

From the above equations

$$x_{k+1} = 13(11 - y_k)$$
$$y_{k+1} = 15(16 - 2x_k + 1)$$

Initial gauss $(x,y)=(0,0)$

1st Approximation

$$x_1 = 3.667$$
$$y_1 = 1.733$$

2nd Approximation

$$x_2 = 3.0889$$
$$y_2 = 1.9644$$

3rd Approximation

$$x_3 = 3.0119$$
$$y_3 = 1.9953$$

4th Approximation

$$x_4 = 3.0016$$
$$y_4 = 1.9999$$

5th Approximation

$$x_5 = 3.0002$$
$$y_5 = 1.9999$$

6*th* Approximation

$$x_6 = 3$$
$$y_6 = 2$$

Solution of the Guass-Seidal method $x = 3, y = 2$

Iterations are tabulated as below

Iteration	x	y
1	3.6667	1.7333
2	3.0889	1.9644
3	3.0119	1.9953
4	3.0016	1.9994
5	3.0002	1.9999
6	3	2

Example.2.5. Apply the Gauss-Seidel iteration method to solve the equations of

$$20x + y - 2z = 17, 3x + 20y - z = -18, 2x - 3y + 20z = 25$$

Solution We write the given equation in the form,

$$x = \frac{1}{20}(17 - y + 2z); y = \frac{1}{20}(-18 - 3x + z); z = \frac{1}{20}(25 - 2x + 3y)$$

We start from the approximation. $x_0 = y_0 = z_0 = 0$.

Substituting $y = y_0, z = z_0$ in the right side of the first of the equations.

We get,

$$x_1 = \frac{1}{20}(17 - y_0 + 2z_0) = 0.8500$$

Putting $x = x_1, z = z_0$ in the 2nd of the equation (i), we obtain

$$y_1 = \frac{1}{20}(-18 - 3x_1 + z_0) = -1.0275$$

Putting $x = x_1, y = y_1$ in the last of the equation (i), we obtain

$$z_1 = \frac{1}{20}(25 - 2x_1 + 3y_1) = 1.0109$$

For the second iteration, we have

$$x_2 = \frac{1}{20}(17 - y_1 + 2z_1) = 1.0025$$

$$y_2 = \frac{1}{20}(-18 - 3x_2 + z_1) = -0.9998$$

$$z_2 = \frac{1}{20}(25 - 2x_2 + 3y_2) = 0.9998$$

For the third iteration, we get

$$x_3 = \frac{1}{20}(17 - y_2 + 2z_2) = 1.0000$$

$$y_3 = \frac{1}{20}(-18 - 3x_3 + z_2) = -1.0000$$

$$z_3 = \frac{1}{20}(25 - 2x_3 + 3y_3) = 1.0000$$

The values in the 2nd and 3rd iterations being practically the same, we can stop.

Hence the solution is $x = 1, y = -1, z = 1$

Example.2.6 Solve the equations using Gauss Seidel method 2x+y+z=5,3x+5y+2z=15,2x+y+4z=8

Solution   Given   2x+y+z=5,3x+5y+2z=15,2x+y+4z=8.

From the above equations

$$x_{k+1} = 12(5 - y_k - z_k)$$
$$y_{k+1} = 15(15 - 3x_k + 1 - 2z_k)$$
$$z_{k+1} = 14(8 - 2x_k + 1 - y_k + 1)$$

Initial gauss $(x,y,z)=(0,0,0)$

1*st* Approximation

$$x_1 = 2.5$$
$$y_1 = 1.5$$
$$z_1 = 0.375$$

2*nd* Approximation

$$x_2 = 1.5625$$
$$y_2 = 1.9125$$
$$z_2 = 0.7406$$

3rd Approximation

$$x_3 = 1.1734$$
$$y_3 = 1.9997$$
$$z_3 = 0.9134$$

4th Approximation

$$x_4 = 1.0435$$
$$y_4 = 2.0086$$
$$z_4 = 0.9761$$

5th Approximation

$$x_5 = 1.0077$$
$$y_5 = 2005$$
$$z_5 = 0.9949$$

6th Approximation

$$x_6 = 1.0001$$
$$y_6 = 2.002$$
$$z_6 = 0.9995$$

7th Approximation

$$x_7 = 0.9993$$
$$y_7 = 2.0007$$
$$z_7 = 1.0002$$

8*th* Approximation

$$x_8 = 0.9996$$
$$y_8 = 2.0002$$
$$z_8 = 1.0002$$

Solution

$$x = 0.9996 \cong 1,$$

$$y = 2.0002 \cong 2,$$
$$z = 1.0002 \cong 1$$

Iterations are tabulated as below

Iteration	x	y	z
1	2.5	1.5	0.375
2	1.5625	1.9125	0.7406
3	1.1734	1.9997	0.9134
4	1.0435	2.0086	0.9761
5	1.0077	2.005	0.9949
6	1.0001	2.002	0.9995
7	0.9993	2.0007	1.0002
8	0.9996	2.0002	1.0002

Example.2.7 Solve equations x+y+z=7,x+2y+2z=13,x+3y+z=13 using Gauss Seidel method.

Solution   Given   x+y+z=7,x+2y+2z=13,x+3y+z=13
The coefficient matrix of the given system is not diagonally dominant.
Hence, we re-arrange the equations as follows, such that the elements in the coefficient matrix are diagonally dominant.

From the above equations

$$x_{k+1} = 11(7 - y_k - z_k)$$
$$y_{k+1} = 13(13 - x_k + 1 - z_k)$$
$$z_{k+1} = 12(13 - x_k + 1 - 2y_k + 1)$$

Initial gauss (x,y,z)=(0,0,0)

Solution steps are

1st Approximation

$$x_1 = 7$$
$$y_1 = 2$$
$$z_1 = 1$$

2nd Approximation

$$x_2 = 4$$
$$y_2 = 2.6667$$
$$z_2 = 1.8333$$

3rd Approximation

$$x_3 = 2.5$$
$$y_3 = 2.8889$$
$$z_3 = 2.3611$$

4th Approximation

$$x_4 = 1.75$$
$$y_4 = 2.963$$
$$z_4 = 2.662$$

5th Approximation

$$x_5 = 1.375$$
$$y_5 = 2.9877$$
$$z_5 = 2.8248$$

6th Approximation

$$x_6 = 1.1875$$
$$y_6 = 2.9959$$
$$z_6 = 2.9104$$

7th Approximation

$$x_7 = 1.0938$$
$$y_7 = 2.9986$$
$$z_7 = 2.9545$$

8th Approximation

$$x_8 = 1.0469$$
$$y_8 = 2.9995$$
$$z_8 = 2.9777$$

9*th* Approximation

$$x_9 = 1.0234$$
$$y_9 = 2.9998$$
$$z_9 = 2.9884$$

10*th* Approximation

$$x_{10} = 1.0117$$
$$y_{10} = 2.9999$$
$$z_{10} = 2.9942$$

11*th* Approximation

$$x_{11} = 1.0059$$
$$y_{11} = 3$$
$$z_{11} = 2.9971$$

12*th* Approximation

$$x_{12} = 1.0029$$
$$y_{12} = 3$$
$$z_{12} = 2.9985$$

13*th* Approximation

$$x_{13} = 1.0015$$
$$y_{13} = 3$$
$$z_{13} = 2.9993$$

14*th* Approximation

$$x_{14} = 1.0007$$
$$y_{14} = 3$$
$$z_{14} = 2.9996$$

Solution of the Guass-Seidal method $x = 1, y = 3, z = 3$
Iterations are tabulated as below

Iteration	x	y	z
1	7	2	1
2	4	2.6667	1.8333
3	2.5	2.8889	2.3611
4	1.75	2.963	2.662
5	1.375	2.9877	2.8248
6	1.1875	2.9959	2.9104
7	1.0938	2.9986	2.9545
8	1.0469	2.9995	2.977
9	1.0234	2.9998	2.9884
10	1.0117	2.9999	2.9942
11	1.0059	3	2.9971

12	1.0029	3	2.9985
13	1.0015	3	2.9993
14	1.0007	3	2.9996

2.3. LU-Decomposition

We start with the linear system

$$Ax = B$$

Possibly the first method that one learns for solving such a linear system of equations is the Gaussian elimination. In computer algebra a slightly modified version of this method is the LU-decomposition, in which one tries to decompose the matrix A according to A = L · U. Knowing the LU decomposition for a matrix A allows us to solve the linear system very easily:

$$Ax = b$$

$$LUx = b$$

$$Ux = L^{-1}b$$

$$x = U^{-1}\left(L^{-1}b\right)$$

Here $L^{-1}b$ uses forward substitution and $U^{-1}\left(L^{-1}b\right)$ backward substitution. Note that sometimes an additional step 'Pivoting', is needed in which either only rows (partial pivoting) or rows and columns (full pivoting) is required, e.g., if you would get zeros on the diagonal. The question arises of how to obtain the LU decomposition? One way uses the recursive leading-row column LU algorithm.

$$\begin{pmatrix} a_{11}a_{12} \\ a_{21}a_{22} \end{pmatrix} = \begin{pmatrix} 1 & 0 \\ I_{21}L_{22} \end{pmatrix} \begin{pmatrix} u_{11}u_{12} \\ 0U_{22} \end{pmatrix}$$

Here it is worth pointing out, if A is an nxn matrix, A22 is a $(n-1)\times(n-1)$-matrix and a12, ... are vectors of length $(n-1)$. The matrix equation can also be rewritten as:

$$a_{11} = u_{11}$$

$$u_{12} = a_{12}$$

$$I_{21} = \frac{1}{u_{11}} a_{21}$$

$$L_{22}U_{22} = A_{22} - a_{21} / (a_{11})^{-1} a_{12}$$

The $(n-1) \times (n-1)$ matrix A22 − a21/(a11) −a12 is the Schur complement and defines a new system of size $(n-1) \times (n-1)$ to solve. Overall, the LU-decomposition has costs proportional to n^3 . Therefore, this method can only be used for 'small' matrices. These can be written in the form of AX = B as:

$$\begin{bmatrix} a_{11} & a_{12} & a_{13} \\ a_{21} & a_{22} & a_{23} \\ a_{31} & a_{32} & a_{33} \end{bmatrix} \begin{bmatrix} x_1 \\ x_2 \\ x_3 \end{bmatrix} = \begin{bmatrix} b_1 \\ b_2 \\ b_3 \end{bmatrix}$$

Here,

$$A = \begin{bmatrix} a_{11} & a_{12} & a_{13} \\ a_{21} & a_{22} & a_{23} \\ a_{31} & a_{32} & a_{33} \end{bmatrix}, X = \begin{bmatrix} x_1 \\ x_2 \\ x_3 \end{bmatrix}, B = \begin{bmatrix} b_1 \\ b_2 \\ b_3 \end{bmatrix}$$

Now follow the steps given below to solve the above system of linear equations by LU Decomposition method.

Step 1: Generate a matrix A = LU such that L is the lower triangular matrix with principal diagonal elements being equal to 1 and U is the upper triangular matrix.

That means,

$$L = \begin{bmatrix} 1 & 0 & 0 \\ l_{21} & 1 & 0 \\ l_{31} & l_{22} & 1 \end{bmatrix} \text{ and } U = \begin{bmatrix} u_{11} & u_{21} & u_{13} \\ 0 & u_{22} & u_{23} \\ 0 & 0 & u_{33} \end{bmatrix}$$

64

Step 2: Now, we can write AX = B as:

$$LUX = B....(1)$$

Step 3: Let us assume $\qquad UX = Y....(2)$

$$\text{Where } Y = \begin{bmatrix} y_1 \\ y_2 \\ y_3 \end{bmatrix}$$

Step 4: From equations (1) and (2), we have;

$$LY = B$$

On solving this equation, we get y_1, y_2, y_3

Step 5: Substituting Y in equation (2), we get UX = Y

By solving equation, we get X, x_1, x_2, x_3.

The above process is also called the Method of Triangularisation.

Example.2.6 Solve the system of equations $x_1 + x_2 + x_3 = 1$, $3x_1 + x_2 - 3x_3 = 5$ and $x_1 - 2x_2 - 5x_3 = 10$ by LU decomposition method.

Solution The given system of equations are:

$$x_1 + x_2 + x_3 = 1$$

$$3x_1 + x_2 - 3x_3 = 5$$

$$x_1 - 2x_2 - 5x_3 = 10$$

These equations are written in the form of AX = B as:

$$\begin{bmatrix} 1 & 1 & 1 \\ 3 & 1 & -3 \\ 1 & -2 & -5 \end{bmatrix}\begin{bmatrix} x_1 \\ x_2 \\ x_3 \end{bmatrix} = \begin{bmatrix} 1 \\ 5 \\ 10 \end{bmatrix}$$

Step 1: Let us write the above matrix as LU = A.

That means,

$$\begin{bmatrix} 1 & 0 & 0 \\ l_{21} & 1 & 0 \\ l_{31} & l_{32} & 1 \end{bmatrix}\begin{bmatrix} u_{11} & u_{12} & u_{13} \\ 0 & u_{22} & u_{23} \\ 0 & 0 & u_{33} \end{bmatrix} = \begin{bmatrix} 1 & 1 & 1 \\ 3 & 1 & -3 \\ 1 & -2 & -5 \end{bmatrix}$$

By expanding the left side matrices, we get;

$$\begin{bmatrix} u_{11} & u_{12} & u_{13} \\ l_{21}u_{11} & l_{21}u_{12}+u_{22} & l_{21}u_{13}+u_{33} \\ l_{31}u_{11} & l_{31}u_{12}+l_{32}u_{22} & l_{31}u_{13}+l_{32}u_{23}+u_{33} \end{bmatrix} = \begin{bmatrix} 1 & 1 & 1 \\ 3 & 1 & -3 \\ 1 & -2 & -5 \end{bmatrix}$$

$$u_{11} = 1, u_{12} = 1, u_{13} = 1$$

$$l_{21}u_{11} = 3,$$

$$l_{21}u_{12} + u_{22} = 1,$$

$$u_{21}u_{13} + u_{23} = -3$$

$$l_{31}u_{11} = 1,$$

$$l_{31}u_{12} + l_{32}u_{22} = -2,$$

$$l_{31}u_{13} + l_{32}u_{23} + u_{33} = -5$$

Solving these equations, we get;

$$u_{22} = -2, u_{23} = -6, u_{33} = 3$$

$$l_{21} = 3, l_{31} = 1, l_{32} = 3/2$$

Step 2: $\qquad\qquad\qquad\qquad$ LUX = B

Step 3: $\qquad\qquad\qquad\qquad$ Let UX = Y

Step 4: From the previous two steps, we have LY = B

Thus,

$$\begin{bmatrix} 1 & 0 & 0 \\ 3 & 1 & 0 \\ 1 & 3/2 & 1 \end{bmatrix} \begin{bmatrix} y_1 \\ y_2 \\ y_3 \end{bmatrix} = \begin{bmatrix} 1 \\ 5 \\ 10 \end{bmatrix}$$

So, $\qquad\qquad\qquad\qquad\qquad y_1 = 1$

$$3y_1 + y_2 = 5$$

$$y_1 + (3/2)\, y_2 + y_3 = 10$$

Solving these equations, we get;

$$y_1 = 1, y_2 = 2, y_3 = 6$$

Step 5: Now, consider UX = Y. So,

$$\begin{bmatrix} 1 & 1 & 1 \\ 0 & -2 & -6 \\ 0 & 0 & 3 \end{bmatrix} \begin{bmatrix} x_1 \\ x_2 \\ x_3 \end{bmatrix} = \begin{bmatrix} 1 \\ 2 \\ 6 \end{bmatrix}$$

By expanding this equation, we get;

$$x_1 + x_2 + x_3 = 1$$

$$-2x_2 - 6x_3 = 2$$

$$3x_3 = 6$$

Solving these equations, we can get;

$$x_3 = 2, x_2 = -7 \text{ and } x_1 = 6$$

Hence, the solution of the given system of equations is (6, -7, 2).

LU Decomposition Method Problems

1. Solve the following equations by LU decomposition method.
   $6x_1 + 18x_2 + 3x_3 = 3$, $2x_1 + 12x_2 + x_3 = 19$, $4x_1 + 15x_2 + 3x_3 = 0$

2. Solve the below given system of equations by LU decomposition.
   $x + y + z = 1$, $4x + 3y - z = 6$, $3x + 5y + 3z = 4$

3. Find the solution of the system of equations by LU decomposition.
   $x + 2y + 3z = 9$, $4x + 5y + 6z = 24$, $3x + y - 2z = 4$

2.4. Tridiagonal system of equations

A system of simultaneous algebraic equations with nonzero coefficients only on the main diagonal, the lower diagonal, and the upper diagonal is called a tridiagonal system of equations. Consider a tridiagonal system of N equations with N unknowns, $u_1, u_2, u_3, ... u_N$ as given below.

A standard method for solving a system of linear, algebraic equations is Gaussian elimination. Thomas' algorithm, also called Tri Diagonal Matrix Algorithm (TDMA) is essentially the result of applying Gaussian elimination to the tridiagonal system of equations. The ith equation in the system may be written as

$$a_i u_{i-1} + b_i u_i + c_i u_{i+1} = d_i$$

where $a_1 = 0$ and $C_N = 0$. Looking at the system of equations, we see that i[th] unknown can be expressed as a function of (i+1) th unknown. That is

$$u_i = P_i u_{i+1} + Q_i$$
$$u_{i-1} = P_{i-1} u_i + Q_{i-1}$$

Where P_i and Q_i are constants. Note that if all the equations in the system are expressed in this fashion, the coefficient matrix of the system would transform to an upper triangular matrix.

To determine the constants P_i and Q_i , we plug equation 4 in 2 to yield

$$a_i P_{i-1} u_i + a_i Q_{i-1} + b_i u_i + c_i u_{i-1} = d_i,$$
$$\left(b_i + a_i P_{i-1} \right) u_i + c_i u_{i+1} = d_i - a_i Q_{i-1}$$
$$u_i = \frac{-c_i}{b_i + a_i P_{i-1}} u_{i+1} + \frac{d_i - a_i Q_{i-1}}{b_i + a_i P_{i-1}}$$

Comparing equations (3) and (5), we obtain

$$P_i = \frac{-c_i}{b_i + a_i P_{i-1}}$$

These are the recurring relations for the constants P and Q. It shows that P_i can be calculated if P_{i-1} is known. To start the computation, we use the fact that $a_1 = 0$. Now, P_1 and Q_1 can be easily calculated because terms involving P_0 and Q_0 vanish. Therefore

$$P_1 = \frac{-c_1}{b_1}, Q_1 = \frac{d_1}{b_1}$$

Once the values of P_1 and Q_1 are known, we can use the recurring expressions for P_i and Q_i for all values of i.

Now, to start the back substitution, We use the fact that $c_N = 0$. As a consequence, from equation (6), we have PN = 0, which results in $u_N = c_N$. Once the value of u_N is known we use equation (3) to obtain $u_{N-1}, u_{N-2}, \ldots u_1$.

2.5. Thomas Algorithm (Tridiagonal system/matrix)

$$A\vec{X} = \vec{b}$$

$$
\begin{bmatrix}
a_{11} & a_{12} & 0 & \cdots & 0 & 0 & 0 & 0 \\
a_{21} & a_{22} & a_{23} & \cdots & 0 & 0 & 0 & 0 \\
0 & a_{32} & a_{33} & a_{34} & 0 & \cdots & 0 & 0 \\
0 & 0 & a_{43} & a_{44} & a_{45} & \cdots & 0 & 0 \\
& & & & & & & \\
0 & 0 & 0 & 0 & 0 & 0 & a_{n\,n-1} & a_{nn}
\end{bmatrix}
\begin{Bmatrix}
X_1 \\ X_2 \\ X_3 \\ \vdots \\ X_n
\end{Bmatrix}
=
\begin{Bmatrix}
b_1 \\ b_2 \\ b_3 \\ \vdots \\ b_n
\end{Bmatrix}
$$

Tridiagonal matrix has 3 non-zero elements in all of its rows, except in the 1st and last row, with one element each on the left and right of the diagonal element. The 1st and last rows have one element on the right and left of the diagonal elements, respectively. It is a banded matrix around its diagonal:

Therefore, a tridiagonal matrix can be represented using single subscripted indices for its elements:

70

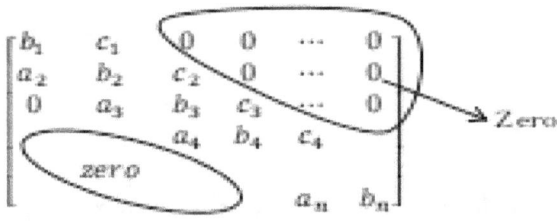

Note that the element in the ith row is represented. All b_s are on the diagonal. Therefore, where it is a tridiagonal matrix and it represents the following set of linear algebraic equations:

$$b_1 x_1 + c_1 x_2 = d_1$$

$$a_2 x_1 + b_2 x_2 + c_2 x_3 = d_2$$

$$a_3 x_2 + b_3 x_3 + c_3 x_4 = d_3$$

$$a_4 x_3 + b_4 x_4 + c_4 x_5 = d_4$$

$$a_n x_{n+1} + b_n x_n = d_n$$

$\Rightarrow$ You must have noted that for the ith row, b_i is the diagonal element multiplied with the variable x_i in the some row, whereas a_i is multiplied with x_{i-1} (the variable i^{th} row) and c_i is multiplied with x_{i+1} (the variable below row). Naturally, the first and last rows have only two variables.

$\Rightarrow$ Tridiagonal system is quite common when using Finite Difference 2nd order method to solve boundary value problems or partial differential.

2.6. Eigen values and Eigen vectors

In the previous unit, you have studied direct methods for solving linear system of equations $Ax = b$, A being $n \times n$ non-singular matrix. Direct methods provide the exact solution in a finite

71

number of steps provided exact arithmetic is used and them is no round-off error. Also, direct methods are generally used when the matrix A is dense or filled, that is, there are few zero elements, and the order of the matrix is not very large say $n < 50$. Iterative methods, on the other hand, start with an initial approximation and by applying suitably chosen algorithm, lead to successively better approximations. Even if the process converges, it would give only an approximate Solution These methods are generally used when the matrix A is sparse and the order of the matrix A is very large say $n > 50$. Sparse matrices have very few non-zero elements. In most cases these non-zero elements lie on or near the main diagonal giving rise to tri-diagonal, five diagonal or band matrix systems. It may be noted that there are no fixed rules to decide when to use direct methods and when to use iterative methods. However, when the coefficient matrix is sparse or large, the use of iterative methods is ideally suited to find the solution which fake advantage of the sparse nature of the matrix involved. Eigen values arise in many physical problems. Flexural oscillations of a tapered rod, buckling of beams, stability of electrical circuits vibration problems etc. are some of the many mas where eigenvalues are required. These problems require the evaluation of a non-zero parameter λ satisfying the matrix equation. $Ax = \lambda x$.

Or $$(A - \lambda I)x = 0$$

The values of λ for which the equation has a non-zero solution are called the **eigenvalues** of the matrix A. Corresponding to these eigenvalues, the non-zero solution of above equation i.e. the vector x are called the **eigenvectors** of A. The problem of finding the eigen values and the corresponding eigenvectors of a square matrix A is known as the **eigenvalue problem.** In this unit we will first discuss two iterative methods, namely, Jacobi iteration and Gauss-Seidel iteration-methods for finding the solution of linear system of equations

$$Ax = b.$$

This will be followed by studying the eigenvalue problem. In the previous three units, we were concerned with the non-homogeneous system of linear equations, $= b$. We know that this system has a unique solution if the matrix A is non-singular. But, if the vector $b = 0$, then the system reduces to the homogeneous system

If the coefficient matrix A, in above equation is non-singular, then system has only the zero solution, $x = 0$. For the homogeneous system to have a nonzero solution, the matrix A must be singular and in this case the solution is not unique.The homogeneous system of above equation will have a nonzero solution only when the coefficient matrix is singular, that is, $\det(A - \lambda I) = 0$.

If the matrix A is an $n \times n$ matrix then equation gives a polynomial of degree n in λ. This polynomial is called the characteristic equation of A. The n roots $\lambda_1, \lambda_2, \ldots, \lambda_n$ of this polynomial are the eigenvalues of A. For each eigenvalue λ_i, there exists a vector x_i (the eigenvector), which is the nonzero solution of the system of equations $(A - \lambda_i)x_i = 0$

The eigenvalues have several exciting properties. We & all now state and prove a few of these properties, which we shall be using frequently

Objectives

After studying this unit, you should be able to:

- obtain the solution of system of linear equations. $Ax = b$ by using iterative methods viz. Jacobi method or the Gauss-Seidel method,
- tell whether the iterative scheme converges or not,

- obtain the value of convergence and the number of iterations needed for required accuracy of the iteration methods,
- solve simple eigenvalue problems,
- calculate largest eigenvalue using power method,
- calculate smallest eigenvalue using inverse power method.

Theorem.2.1 A matrix A is singular if and only if it has zero eigenvalue.

Proof. If A has a zero eigen value then

$$\det(A - 0I) = 0$$

$$\Rightarrow \det(A) = 0$$

$$\Rightarrow A \text{ is singular.}$$

Conversely, if A is singular then

$$\det(A) = 0$$

$$\Rightarrow \det(A - 0I) = 0$$

$$\Rightarrow 0 \text{ is an eigen value of the matrix } A.$$

Theorem. 2.2 A and A^T have the same eigenvalues.

Proof. If λ is an eigenvalue of A then

$$\det(A - \lambda I) = 0$$

$$\Rightarrow \det(A - \lambda I)^T = 0$$

$$\Rightarrow \det(A^T - \lambda I)^T = 0$$

$$\Rightarrow \det(A^T - \lambda I) = 0$$

$$\Rightarrow \text{ is an eigen value of } A^T$$

74

However, the eigen vectors of A and A^T are not the same.

Theorem.2.3 If the eigenvalues of a matrix A are $\lambda_1, \lambda_2, \ldots, \lambda_n$ then the eigenvalues of A^m, m any positive integer, are $\lambda_1^m, \lambda_2^m, \ldots, \lambda_n^m$. Also both the matrices A and A^m have the same set of eigenvectors.

Proof. Since $\lambda_i (i = 1, 2, \ldots, n)$ are the eigenvalues of A, we have

$$Ax = \lambda_i x^2, i = 1, 2, \ldots, n$$

Pre-multiplying Equation by A on both sides, we get

$$A^2 x = A \lambda_i x = \lambda_i (Ax) = \lambda_i^2 x$$

which implies that $\lambda_1^2, \lambda_2^2, \ldots, \lambda_n^2$, are the eigenvalues of A^2. Further, A and A^2 have the same eigenvectors. Pre-multiplying equation $(m - 1)$ times by A on both sides the general result follows.

Theorem.2.4 If $\lambda_1, \lambda_2, \ldots, \lambda_n$ are the eigenvalues of A, then $1/\lambda_1, 1/\lambda_2, \ldots, 1/\lambda_n$ are the eigenvalues of A^{-1}. Also, both the matrices A and A^{-1} have the same set of eigenvectors.

Proof. Since, $\lambda_i (i = 1, 2, \ldots, n)$ are the eigenvalues of A, we have

$$Ax = \lambda_i x, i = 1, 2, \ldots, n$$

Pre-multiplying Equation on both sides by A^{-1}, we get

$$A^{-1} Ax = \lambda_i A^{-1} x$$

which gives

$$x = \lambda_i A^{-1} x$$

$$\text{or} \quad A^{-1} x = \frac{1}{\lambda_i} x$$

75

Theorem.2.5. If $\lambda_1, \lambda_2, \ldots, \lambda_n$ are the eigenvalues of A, then $\lambda_i - q, i = 1, 2, \ldots, n$ are the eigenvalues of $A - qI$ for any real number q. Both the matrices A and $A - qI$ have the same set of eigen vectors.

Proof.   Since λ_i is an eigen values of A, we have

$$Ax = \lambda_i x, i = 1, 2, \ldots, n$$

Subtracting qx from both sides of Eqn. (11.40), we get

$$Ax - qx = \lambda_i x - qx$$

which gives $(A - qI)x = (\lambda_i - q)$

Theorem.2.6. If $\lambda_i, i = 1, 2, \ldots, n$ are the eigenvalues of A then $\frac{1}{\lambda_i - q}, i = 1, 2, \ldots, n$ are the eigen values of $(A - qI)^{-1}$ for any real number q. Both the matrices A and $(A - qI)^{-1}$ have the same set of eigenvectors.

. We leave the proof to you.

We now give you a direct method of calculating the eigenvalues and eigenvectors of a matrix.

Example.2.7 Find the eigen values of the matrix

$$a) \; A = \begin{bmatrix} 1 & 0 & 0 \\ 0 & 2 & 0 \\ 0 & 0 & 3 \end{bmatrix} b) \; A = \begin{bmatrix} 1 & 0 & 0 \\ 2 & 3 & 0 \\ 4 & 5 & 6 \end{bmatrix} c) \; A = \begin{bmatrix} 1 & 2 & 3 \\ 0 & 4 & 5 \\ 0 & 0 & 6 \end{bmatrix}$$

Solution: a) Using equation, we obtain the characteristic equations as

$$\det(A - \lambda I) = \begin{vmatrix} 1 - \lambda & 0 & 0 \\ 0 & 2 - \lambda & 0 \\ 0 & 0 & 3 - \lambda \end{vmatrix} = 0$$

76

which gives $(1 - \lambda)(2 - \lambda)(3 - \lambda) = 0$.

the eigen values of A are $\lambda_1 = 1, \lambda_2 = 2, \lambda_3 = 3$.

b) $\det(A - \lambda I) = \begin{vmatrix} 1 - \lambda & 0 & 0 \\ 2 & 3 - \lambda & 0 \\ 4 & 5 & 6 - \lambda \end{vmatrix} = 0$

which gives $(1 - \lambda)(3 - \lambda)(6 - \lambda) = 0$.

Eigen values of A are $\lambda_1 = 1, \lambda_2 = 3, \lambda_3 = 6$.

c) $\det(A - \lambda I) = \begin{vmatrix} 1 - \lambda & 2 & 3 \\ 0 & 4 - \lambda & 5 \\ 0 & 0 & 6 - \lambda \end{vmatrix} = 0$

Therefore, $(1 - \lambda)(4 - \lambda)(6 - \lambda) = 0$.

Eigen values of A are $\lambda_1 = 1, \lambda_2 = 4, \lambda_3 = 6$.

Remark: Observe that in example, the matrix , A is diagonal and in parts (b) and (c), it is lower and upper triangular respectively. In these case the eigenvalues of , A are the diagonal elements. This is true for any diagonal, lower trmgular or upper triangular matrix. Formally, we give the result in the following theorem.

Theorem 2.7 The eigenvalues of a diagonal, lower triangular or an upper triangular matrix are the diagonal elements themselves. Let us consider another example.

Example.2.8 Find the eigenvalues and the corresponding eigenvectors of the matrices.

$$a) \begin{bmatrix} 2 & 2 \\ 1 & 3 \end{bmatrix}; \quad b) \begin{bmatrix} 1 & 2 \\ 0 & 1 \end{bmatrix}; \quad c) \begin{bmatrix} 1 & -2 \\ 2 & 1 \end{bmatrix}$$

Solution a) Using Equation, we obtain the characteristic equation

$$|A - \lambda I| = \begin{vmatrix} 2 - \lambda & 2 \\ 1 & 3 - \lambda \end{vmatrix} = 0$$

which gives $\lambda^2 - 5\lambda + 4 = 0$

$$(\lambda - 1)(\lambda - 4) = 0$$

The matrix A has two distinct real eigenvalues $\lambda_1 = 1, \lambda_2 = 4$. To obtain the wrresponding eigenvectors we solve the system of Eqns. for each value of λ.

For $\lambda = 1$, we obtain the system of equations

$$x_1 + 2x_2 = 0$$
$$x_1 + 2x_2 = 0$$

which reduces to a single equation

$$x_1 + 2x_2 = 0$$

Taking $x_2 = k$, we get $x_1 = -2k$, k being arbitraty nonzen, constant. Thus, the eigenvector is of the form

$$\begin{bmatrix} x_1 \\ x_2 \end{bmatrix} = \begin{bmatrix} -2k \\ k \end{bmatrix} = k \begin{bmatrix} -2 \\ 1 \end{bmatrix}$$

For $\lambda = 4$, we obtain the system of equations

$$-2x_1 + 2x_2 = 0$$
$$x_1 - x_2 = 0$$

which reduces to a single equation

$$x_1 - x_2 = 0$$

Taking $x_2 = k$, we get $x_1 = k$ and the corresponding eigenvector is

$$\begin{bmatrix} x_1 \\ x_2 \end{bmatrix} = k \begin{bmatrix} 1 \\ 1 \end{bmatrix}$$

Note: In practice we usually omit k and say that $[-2 \ 1]^T$ and $[1 \ 1]^T$ are the eigenvectors of A-aqresponding to the eigenvalues $\lambda = 1$ and $\lambda = 4$ respectively. Moreover, the eigenvectors in this case are **linearly independent.**

The characteristic equation in this case becomes

$$(\lambda - 1)^2 = 0.$$

Taking $x_1 = k$, we obtain the eigenvector as

$$\begin{bmatrix} x_1 \\ x_2 \end{bmatrix} = k \begin{bmatrix} 1 \\ 0 \end{bmatrix}$$

Note: that, in this case of repeated eigenvalues, we got linearly dependent eigenvectors.

The characteristic equation in this case becomes

$$\lambda^2 - 2\lambda + 5 = 0$$

which gives two complex eigenvalues $\lambda = 1 \pm 2i$.

The eigenvector corresponding to $\lambda = 1 \pm 2i$ is the solution of the system of Eqns. (11.36). In this case, we obtain the following equations

$$ix_1 + x_2 = 0$$

$$x_1 - ix_2 = 0$$

which reduces to a single equation

$$x_1 - ix_2 = 0$$

Taking $x_2 = k$, we get the eigenvector

$$\begin{bmatrix} x_1 \\ x_2 \end{bmatrix} = k \begin{bmatrix} -i \\ 1 \end{bmatrix}$$

Similarly, for $\lambda = 1 - 2i$, we obtain the eigenvector

$$\begin{bmatrix} x_1 \\ x_2 \end{bmatrix} = k \begin{bmatrix} -i \\ 1 \end{bmatrix}$$

In the above problem, you may note that we got complex eigenvectors corresponding to complex eigenvalues. Let us now consider an example of a 3 × 3 matrix.

Remark: The characteristic equation of the square matrix A is given by $\det(A - \lambda I) = 0$. For a matrix of order 3, the following formula is helpful for getting a cubic polynomial in λ without even expanding the determinant directly.

λ^3 − (sum of the diagonal elements of A) λ^2 + (sum of the minors of the diagonal elements of A) $\lambda - \det(A) = 0$.

Example.2.9 Determine the eigenvalues and the corresponding eigenvectors for the matrices

$$a) \begin{bmatrix} 2 & -1 & 0 \\ -1 & 2 & -1 \\ 0 & -1 & 2 \end{bmatrix}; \quad b)\ A = \begin{bmatrix} 6 & -2 & 2 \\ -2 & 3 & -1 \\ 2 & -1 & 3 \end{bmatrix}$$

Solution The characteristic equation in this case becomes

$$\begin{vmatrix} 2-\lambda & -1 & 0 \\ -1 & 2-\lambda & -1 \\ 0 & -1 & 2-\lambda \end{vmatrix} = 0$$

which using the above remark gives the polynomial
$$\lambda^3 - (2 + 2 + 2)\lambda^2 + (3 + 4 + 3)\lambda - 4$$

$$= \lambda^3 - 6\lambda^3 - 10\lambda - 4$$

$$= (2 - \lambda)(\lambda^2 - 4\lambda + 2) = 0$$

Therefore, the eigenvector of A are $2, 2 + \sqrt{2}$ and $2 - \sqrt{2}$.

The eigenvector of A corresponding to $\lambda = 2$ is the solution of the system

of Equations, which reduces to

$$x_2 = 0$$

$$x_1 + x_3 = 0$$

Taking $x_3 = k$, we obtain the eigenvector

$$\begin{bmatrix} x_1 \\ x_2 \\ x_3 \end{bmatrix} = k \begin{bmatrix} -1 \\ 0 \\ 1 \end{bmatrix}$$

The eigenvector of A corresponding to $\lambda = 2 + \sqrt{2}$ is the solution of the system of equations

$$\begin{bmatrix} -\sqrt{2} & -1 & 0 \\ -1 & -\sqrt{2} & -1 \\ 0 & -1 & -\sqrt{2} \end{bmatrix} \begin{bmatrix} x_1 \\ x_2 \\ x_3 \end{bmatrix} = \begin{bmatrix} 0 \\ 0 \\ 0 \end{bmatrix}$$

To find the solution of system of equations, we use Gauss elimination method.

Performing $R_2 - \frac{1}{\sqrt{2}} R_1$, we get

$$\begin{bmatrix} -\sqrt{2} & -1 & 0 \\ -1 & -1/\sqrt{2} & -1 \\ 0 & -1 & -\sqrt{2} \end{bmatrix} \begin{bmatrix} x_1 \\ x_2 \\ x_3 \end{bmatrix} = \begin{bmatrix} 0 \\ 0 \\ 0 \end{bmatrix}$$

Again performing $R_3 - \sqrt{2} R_2$ we get

$$\begin{bmatrix} -\sqrt{2} & -1 & 0 \\ 0 & -1/\sqrt{2} & -1 \\ 0 & 0 & 0 \end{bmatrix} \begin{bmatrix} x_1 \\ x_2 \\ x_3 \end{bmatrix} = \begin{bmatrix} 0 \\ 0 \\ 0 \end{bmatrix}$$

which give the equations

$$-\sqrt{2}x_1 - x_2 = 0 \text{ and } -x_2 - \sqrt{2}x_3 = 0$$

Similarly, corresponding to the eigenvalue $\lambda = 2 - \sqrt{2}$ the eigenvector is the solution of system of equations

$$\begin{bmatrix} -\sqrt{2} & -1 & 0 \\ -1 & -\frac{1}{\sqrt{2}} & -1 \\ 0 & -1 & -\sqrt{2} \end{bmatrix} \begin{bmatrix} x_1 \\ x_2 \\ x_3 \end{bmatrix} = \begin{bmatrix} 0 \\ 0 \\ 0 \end{bmatrix}$$

Using the Gauss elimination method, the system reduces to the equations

$$\sqrt{2}x_1 - x_2 = 0$$

$$x_2 - \sqrt{2}x_3 = 0$$

Taking $x_3 = k$, we obtain the eigenvector

$$\begin{bmatrix} x_1 \\ x_2 \\ x_3 \end{bmatrix} = k \begin{bmatrix} 1 \\ \sqrt{2} \\ 1 \end{bmatrix}$$

a) The characteristic equation in this case becomes
$$(\lambda - 8)(\lambda - 2)^2 = 0$$

Therefore, the matrix A has the real eigenvalues 8,2 and 2. The eigenvalue 2 is repeated two times.

The eigenvector corresponding to $\lambda = 8$ is solution of system of Equations, which reduces to

$$x_1 + x_2 - x_3 = 0$$
$$2x_1 + 5xx_2 - x_3 = 0$$
$$2x_1 - x_2 - 5x_3 = 0$$

Subtracting the last equation of system from the second equation we obtain the system of equations

$$x_1 + x_2 - x_3 = 0$$
$$x_2 + x_3 = 0$$

Taking $x_3 = k$, the eigenvector is

$$\begin{bmatrix} x_1 \\ x_2 \\ x_3 \end{bmatrix} = k \begin{bmatrix} 2 \\ -1 \\ 1 \end{bmatrix}$$

The eigenvector corresponding to $\lambda = 2$ is the solution of system of equations, which reduces to a single equation. $2x_1 - x_2 + x_3 = 0$

We can take any values for x_1, and x_2, which need not be related to each other. The two linearly independent solutions can be written as:

$$k \begin{bmatrix} 1 \\ 0 \\ -2 \end{bmatrix} \text{ or } k \begin{bmatrix} 0 \\ 1 \\ 1 \end{bmatrix}$$

2.7. The Jacobi iteration method

We write the system of equations in the form

$$a_{11}x_1 + a_{12}x_2 + \cdots + a_{1n}x_n = b_1$$

$$a_{21}x_1 + a_{22}x_2 + \cdots + a_{2n}x_n = b_2$$

$$\cdot \qquad \cdot \qquad \qquad \cdot \qquad \cdot$$

$$\cdot \qquad \cdot \qquad \qquad \cdot \qquad \cdot$$

$$\cdot \qquad \cdot \qquad \qquad \cdot \qquad \cdot$$

$$a_{n1}x_1 + a_{n2}x_2 + \cdots + a_{nn}x_n = b_n$$

We assume that $a_{11}, a_{22}, \ldots a_{nn}$, are pivot elements and $a_{ii} \neq 0, i = 1,2, \ldots, n$. If any of the pivots is zero, we can interchange the equations to obtain non-zero pivots (partial pivoting).

Note: A being a non-singular matrix, it is possible for us to make all the pivots non-zero. It is only when the matrix A is singular that even complete pivoting may not lead to all the non-zero pivots.

We rewrite system equations in the form and define the

Jacobi iteration method as

$$x_1^{(k+1)} = -\frac{1}{a_{11}}\left(a_{12}x_2^{(k)} + a_{13}x_3^{(k)} + \cdots a_{1n}x_n^{(k)} - b_1\right)$$

$$x_2^{(k+1)} = -\frac{1}{a_{22}}\left(a_{21}x_2^{(k)} + a_{23}x_3^{(k)} + \cdots a_{2n}x_n^{(k)} - b_2\right)$$

and so on

$$x_n^{(k+1)} = -\frac{1}{a_{nn}}\left(a_{n1}x_1^{(k)} + a_{n2}x_2^{(k)} + \cdots a_{n,n-1}x_{n-1}^{(k)} - b_n\right)$$

In the abridged notation, the above set of equations can be written as

$$x_i^{(k+1)} = -\frac{1}{a_{ii}}\left[\sum_{\substack{j=1 \\ i \neq j}}^{n} a_{ij}\,x_j^{(k)} - b_1\right], i = 1,2,\dots,n,$$

$$k = 0,1,..$$

The method can be put in the matrix form as

$$\begin{bmatrix} x_1^{(k+1)} \\ x_2^{(k+1)} \\ \cdot \\ \cdot \\ \cdot \\ x_n^{(k+1)} \end{bmatrix} = -\begin{bmatrix} \frac{1}{a_{11}} & & & \\ & \frac{1}{a_{22}} & & \\ & & \ddots & \\ & & & \frac{1}{a_{nn}} \end{bmatrix}\left\{\begin{bmatrix} 0 & a_{12} & \dots & a_{1n} \\ a_{21} & 0 & \dots & a_{2n} \\ \cdot & & & \\ \cdot & & & \\ a_{n1} & a_{n2} & \dots & 0 \end{bmatrix}\begin{bmatrix} x_1^{(k)} \\ x_2^{(k)} \\ \cdot \\ \cdot \\ x_n^{(k)} \end{bmatrix} - \begin{bmatrix} b_1 \\ b_2 \\ \cdot \\ \cdot \\ b_n \end{bmatrix}\right\}$$

$$x^{(k+1)} = -D^{-1}(L + U)x^{(k)} + D^{-1}b, k = 0,1,\dots$$

where,

$$D = \begin{bmatrix} a_{11} & 0............0 \\ 0 & a_{22}.........0 \\ 0...................a_{nn} \end{bmatrix}, L = \begin{bmatrix} 0 & 0................0 \\ a_{21} & 0.................0 \\ a_{31} & a_{32} & 0..........0 \\ .... \\ a_{n1} & a_{n2} & a_{nn-1}. & 0 \end{bmatrix}$$

$$U = \begin{bmatrix} 0 & a_{12} & a_{13} .... & a_{1n} \\ 0 & 0 & a_{23} & a_{2n} \\ . & & .... \\ . & . & & \\ . & . & & a_{n-1, n.} \\ 0 & 0 & 0......0 \end{bmatrix}$$

The method is of the form,

$$H = -D^{-1}(L + U) \text{ and } c = D^{-1}b$$

Using an equation for computational purposes, we obtain the solution vector $xk+1$ at the $(k+1)$th iteration, element by element. For large n, we rarely use the method in its matrix form as given by the equation.

In this method, in the $(k+1)$th iteration, we use the values, obtained at the *kth* iteration viz., $x1k, x2k, ..., xnk$ on the right-hand side of the equation and get the solution vector $x^{(k+1)}$. We then replace the entire vector $x^{(k)}$ on the right side of the equation by $x^{(k+1)}$ to obtain the solution at the next iteration. In other words, each equation is simultaneously changed by using the method's recent set of x-values. For this reason, this method is also known as the method of simultaneous displacement or **corrections.**

Let us now solve a few examples for better understanding of the method and its convergence

Example 2.10 Perform four iterations of the Jacobi method for solving the system of equations.

$$\begin{bmatrix} -8 & 1 & 1 \\ 1 & -5 & 1 \\ 1 & 1 & -4 \end{bmatrix} \begin{bmatrix} x_1 \\ x_2 \\ x_3 \end{bmatrix} = \begin{bmatrix} 1 \\ 16 \\ 7 \end{bmatrix}$$

with $x^{(0)} = 0$, the exact solution is $x = [-1 \quad -4 \quad -3]^T$.

Solution The Jacobi method, when applied to the system of Equations, becomes

$$x_1^{(k+1)} = \frac{1}{8}\left[x_2^{(k)} + x_3^{(k)} - 1\right]$$

$$x_2^{(k+1)} = \frac{1}{5}\left[x_1^{(k)} + x_3^{(k)} - 16\right]$$

$$x_3^{(k+1)} = \frac{1}{4}\left[x_1^{(k)} + x_2^{(k)} - 7\right], k = 0,1,..$$

Starting with $x^{(0)} = [0 \quad 0 \quad 0]^T$, we obtain from Eqns (11.17), the following results:

$$k = 0$$

$$x_1^{(1)} = \frac{1}{8}[0 + 0 - 1] = -0.125$$

$$x_2^{(1)} = \frac{1}{5}[0 + 0 - 16] = -3.2$$

$$x_3^{(1)} = \frac{1}{4}[0 + 0 - 7] = -1.75$$

$$k = 1$$

$$x_1^{(2)} = \frac{1}{8}[-3.2 - 1.75 - 1] = -0.7438$$

86

$$x_2^{(2)} = \frac{1}{5}[-0.125 - 1.75 - 16] = -3.5750$$

$$x_3^{(2)} = \frac{1}{4}[-0.125 - 3.2 - 7] = -2.5813$$

$$k = 2$$

$$x_1^{(3)} = \frac{1}{8}[-3.5750 - 2.5813 - 1] = -0.8945$$

$$x_2^{(3)} = \frac{1}{5}[-0.7438 - 2.5813 - 16] = -3.8650$$

$$x_3^{(3)} = \frac{1}{4}[-0.7438 - 3.5750 - 7] = -2.8297$$

$$k = 3$$

$$x_1^{(4)} = \frac{1}{8}[-3.8650 - 2.8297 - 1] = -0.9618$$

$$x_2^{(4)} = \frac{1}{5}[-0.8945 - 2.8297 - 16] = -3.9448$$

$$x_3^{(4)} = \frac{1}{4}[-0.8945 - 3.8650 - 7] = -2.9399$$

Thus, after four iterations we get the solution as given in Equations. We find that after each iteration, we get better approximation to the exact Solution

Example 2.11 Jacobi method is used to solve the system of equations

$$\begin{bmatrix} 4 & -1 & 1 \\ 4 & -8 & 1 \\ -2 & 1 & 5 \end{bmatrix} \begin{bmatrix} x_1 \\ x_2 \\ x_3 \end{bmatrix} = \begin{bmatrix} 7 \\ -21 \\ 15 \end{bmatrix}$$

Determine the rate of convergence of the method and the number of iterations needed to make max $\left| \varepsilon_i^{(k)} \right| \leq 10^{-2}$.

Perform these number of iterations starting with initial approximation $x^{(0)} = \begin{bmatrix} 1 & 2 & 2 \end{bmatrix}^T$ and compare the result with the exact solution $\begin{bmatrix} 2 & 4 & 3 \end{bmatrix}^T$.

Solution The Jacobi method when applied to the system of Eqns, gives the iteration matrix

$$H = -\begin{bmatrix} \dfrac{1}{a_{11}} & 0 & 0 \\ 0 & \dfrac{1}{a_{22}} & 0 \\ 0 & 0 & \dfrac{1}{a_{33}} \end{bmatrix} \begin{bmatrix} 0 & a_{12} & a_{13} \\ a_{21} & 0 & a_{23} \\ a_{31} & a_{32} & 0 \end{bmatrix}$$

$$= -\begin{bmatrix} \dfrac{1}{4} & 0 & 0 \\ 0 & -\dfrac{1}{8} & 0 \\ 0 & 0 & \dfrac{1}{5} \end{bmatrix} \begin{bmatrix} 0 & -1 & 1 \\ 4 & 0 & 1 \\ -2 & 1 & 0 \end{bmatrix}$$

$$= \begin{bmatrix} 0 & \dfrac{1}{4} & -\dfrac{1}{4} \\ \dfrac{1}{2} & 0 & \dfrac{1}{8} \\ \dfrac{2}{5} & -\dfrac{1}{5} & 0 \end{bmatrix}$$

The eigenvalue of the iteration matrix H are the roots of the characteristic equation

$$\det(H - \lambda I) = 0$$

Now, $\quad \text{Det}(H - \lambda I) = \begin{vmatrix} -\lambda & \dfrac{1}{4} & -\dfrac{1}{4} \\ \dfrac{1}{2} & -\lambda & \dfrac{1}{8} \\ \dfrac{2}{5} & -\dfrac{1}{5} & -\lambda \end{vmatrix} = \lambda^3 - \dfrac{3}{80} = 0$

All the three eigenvalues of the matrix H are equal and they are equal to

$$\lambda = 0.3347.$$

The spectral radius is

$$R(H) = 0.3347$$

We obtain the rate of convergence

$$v = -\log_{10}(0.3347) = 0.753$$

The number of iterations needed for the required accuracy is given by

$$k = \frac{2}{v} = 5$$

The Jacobi method when applied to the system of Equations, becomes

$$x^{(k+1)} = \begin{bmatrix} 0 & \frac{1}{4} & -\frac{1}{4} \\ \frac{1}{2} & 0 & \frac{1}{8} \\ \frac{2}{5} & -\frac{1}{5} & 0 \end{bmatrix} x^{(k)} + \begin{bmatrix} \frac{7}{4} \\ -\frac{21}{8} \\ 3 \end{bmatrix}, k = 0,1, \ldots$$

starting with the initial approximation $x^{(0)} = [0 \quad 0 \quad 0]^T$, we get from Equation

$$x^{(1)} = [1.75 \quad 3.375 \quad 3.0]^T$$

$$x^{(2)} = [1.8437 \quad 3.875 \quad 3.025]^T$$

$$x^{(3)} = [1.9625 \quad 3.925 \quad 3.9625]^T$$

$$x^{(4)} = [1.9906 \quad 3.9766 \quad 3.0000]^T$$

$$x^{(5)} = [1.9941 \quad 3.9953 \quad 3.0009]^T$$

which is the result after five iterations. Thus, you can see that result obtained after five iterations is quite close to the exact solution $[2 \quad 4 \quad 3]^T$.

Example.2.12 Perform four iterations of the Jacobi method for solving the system of equations.

$$\begin{bmatrix} 2 & -1 & 0 & 0 \\ -1 & 2 & -1 & 0 \\ 0 & -1 & 2 & -1 \\ 0 & 0 & -1 & 2 \end{bmatrix} \begin{bmatrix} x_1 \\ x_2 \\ x_3 \\ x_4 \end{bmatrix} = \begin{bmatrix} 1 \\ 0 \\ 0 \\ 1 \end{bmatrix}$$

with $x^{(0)} = [0.5 \quad 0.5 \ 0.5 \quad 0.5]^T$ .What can you say about the solution obtained if the exact solution is $x = [1 \quad 1 \ 1 \quad 1]^T$?

Solution The Jacobi method when applied to the system of equations, becomes

$$x_1^{(k+1)} = \frac{1}{2}\left[1 + x_2^{(k)}\right]$$

$$x_2^{(k+1)} = \frac{1}{2}\left[x_1^{(k)} + x_3^{(k)}\right]$$

$$x_3^{(k+1)} = \frac{1}{2}\left[x_2^{(k)} + x_4^{(k)}\right]$$

$$x_4^{(k+1)} = \frac{1}{2}\left[1 + x_3^{(k)}\right], k = 0,1,..$$

Using $\quad x^{(0)} = [0.5 \quad 0.5 \ 0.5 \quad 0.5]^T$, we obtain

$$x^{(1)} = [0.75 \quad 0.5 \ 0.5 \quad 0.75]^T$$

$$x^{(2)} = [0.75 \quad 0.625 \ 0.625 \quad 0.75]^T$$

$$x^{(3)} = [0.8125 \quad 0.6875 \ 0.6875 \quad 0.8125]^T$$

$$x^{(4)} = [0.8438 \quad 0.75 \ 0.75 \quad 0.8438]^T$$

You may notice here that the solution is improving after each iteration. Also, the solution obtained after four iterations is not a good approximation to the exact solution $= [1 \quad 1 \ 1 \quad 1]^T$.

This shows that we require a few more iterations to get a I good approximation.

Example.2.13 Find the spectral radius of the iteration matrix when the Jacobi method, is applied to the system of equations.

$$\begin{bmatrix} 1 & 0 & 2 \\ 0 & 1 & -2 \\ 1 & -1 & 1 \end{bmatrix} \begin{bmatrix} x_1 \\ x_2 \\ x_3 \end{bmatrix} = \begin{bmatrix} -1 \\ 5 \\ -3 \end{bmatrix}$$

Verify that the iterations do not converge to the exact solution $x = [1 \quad 3 \quad -1]^T$.

Solution The iteration matrix H in this case becomes

$$H = - \begin{bmatrix} 1 & 0 & 0 \\ 0 & 1 & 0 \\ 0 & 0 & 1 \end{bmatrix} \begin{bmatrix} 0 & 0 & 2 \\ 0 & 0 & -2 \\ 1 & -1 & 0 \end{bmatrix} = \begin{bmatrix} 0 & 0 & -2 \\ 0 & 0 & 2 \\ -1 & 1 & 0 \end{bmatrix}$$

$$c = [-1 \quad 5 \quad -3]^T.$$

The eigenvalue of H are the roots of the characteristic equation

$$\det(H - \lambda I) = 0.$$

This gives us $-\lambda(\lambda^2 - 4) = 0$

i.e., $\lambda = 0, \pm 2$

$\therefore R(H) = 2 > 1$

Thus, the condition in Theorem 1 is violated. The iteration method does not converge.

We now perform few iterations and see what happens actually. Taking $x^{(0)} = 0$ and using the Jacobi method

$$x^{(k+1)} = \begin{bmatrix} 0 & 0 & -2 \\ 0 & 0 & 2 \\ -1 & 1 & 0 \end{bmatrix} x^{(k)} + \begin{bmatrix} -1 \\ 5 \\ -3 \end{bmatrix}$$

We obtain

$$x^{(1)} = [-1 \quad 5 \quad -3]^T$$

$$x^{(2)} = [5 \quad -1 \quad 3]^T$$

$$x^{(3)} = [-7 \quad 11 \quad -9]^T$$

$$x^{(4)} = [17 \quad -13 \quad 15]^T$$

$$x^{(5)} = [-31 \quad 35 \quad -33]^T$$

and so on, which shows that the iterations are diverging fast, you may also try to obtain the solution with other initial approximations.

Let us now consider an example to show that the convergence criterion given in Theorem 2 is only a sufficient condition. That is, there are system of equations which are not diagonally dominant but, the Jacobi iteration method converges.

Example.2.14 Perform iterations of the Jacobi method for solving the system of equations

$$\begin{bmatrix} 1 & 1 & 1 \\ 0 & 2 & 0 \\ 0 & 3 & -1 \end{bmatrix} \begin{bmatrix} x_1 \\ x_2 \\ x_3 \end{bmatrix} = \begin{bmatrix} 3 \\ 2 \\ 1 \end{bmatrix}$$

with $x^{(0)} = [0 \quad 1 \quad 1]^T$. What can you say about the solution obtained if the exact solution is $x = [0 \quad 1 \quad 2]^T$?

Solution The Jacobi method, when applied to the given system of equations, becomes

$$x_1^{(k+1)} = \left[3 - x_2^{(k)} - x_3^{(k)}\right]$$
$$x_2^{(k+1)} = 1$$
$$x_3^{(k+1)} = \left[-1 + 3x_2^{(k)}\right], k = 0,1, \dots$$

Using
$$x^{(0)} = [0 \quad 1 \quad 1]^T,$$
$$x^{(1)} = [1 \quad 1 \quad 2]^T$$
$$x^{(2)} = [0 \quad 1 \quad 2]^T$$

92

$$x^{(3)} = [0 \quad 1 \quad 2]^T$$

You may notice here that the coefficient matrix is not diagonally dominant but the iterations converge to the exact solution after only two iterations.

Example Perform five iterations of the Jacobi method for solving the system of equations given with $x^{(0)} = [1 \quad 1 \quad 1]^T$.

Perform four iterations of the Jacobi method for solving the system of equations.

$$\begin{bmatrix} 5 & 2 & 2 \\ 2 & 5 & 3 \\ 2 & 1 & 5 \end{bmatrix} \begin{bmatrix} x_1 \\ x_2 \\ x_3 \end{bmatrix} = \begin{bmatrix} 1 \\ -6 \\ -4 \end{bmatrix}$$

with $x^{(0)} = 0$. Exact solution is $x = [1 \quad -1 \quad -1]^T$.

Example Perform four iterations of the Jacobi method for solving the system of equations.

$$\begin{bmatrix} 2 & -\frac{1}{2} & 0 \\ -\frac{3}{2} & 2 & -\frac{1}{2} \\ 0 & -\frac{3}{2} & 2 \end{bmatrix} \begin{bmatrix} x_1 \\ x_2 \\ x_3 \end{bmatrix} = \begin{bmatrix} \frac{3}{2} \\ 0 \\ \frac{1}{2} \end{bmatrix}$$

with $x^{(0)} = 0$. Exact solution is $x = [1 \quad 1 \quad 1]^T$.

2.8. Power method

Let us consider the eigenvalue problem

$$Ax = \lambda x.$$

Let $\lambda_1, \lambda_2, \ldots, \lambda_n$, be the n **real** and **distinct eigenvalues** of A such that

$$|\lambda_1| > |\lambda_2| > \cdots > |\lambda_n|$$

Therefore, λ_1, is the dominant eigenvalue of A.

The Power method is one of the simplest methods of finding simultaneously the eigenvalue and vector of a matrix A. This is an iterative method and is suitable in problems in which only a very few roots are required.

In this method, we start with an arbitrary nonzero vector $y^{(0)}$ (not an eigenvector), and form a sequence of vectors $(y^{(k)})$ given by the relation

$$y^{(k+1)} = Ay^{(k)}, k = 0,1,...$$

In the limit as $k \to \infty, y^{(k)}$ converges to the eigenvector corresponding to the dominant eigenvalue of the matrix A. We can stop the iteration when the largest element in magnitude in $y^{(k+1)} - y^{(k)}$ is less than the predefined error tolerance. For simplicity, we usually take the initial vector $y^{(0)}$ with all its elements equal to one.

Note: that multiplying the matrix A with the vector $y^{(k)}$, the elements of the vector $y^{(k+1)}$ may become very large. To avoid this, we normalise (or scale) the vector $y^{(k)}$ at each step by dividing $y^{(k)}$, by its largest element in magnitude. This will make the largest element in magnitude in the vector $y^{(k+1)}$ as one and the remaining elements less than one.

1f $y^{(k)}$ represents the unscaled vector and $v^{(k)}$ the scaled vector then, we have the power method.

$$y^{(k+1)} = Av^{(k)}$$

$$v^{(k+1)} = \frac{1}{m_{k+1}} y^{(k+1)}, k = 0,1,...$$

with, $v^{(0)} = y^{(0)}$ and m_{k+1}, being the largest element in magnitude of $y^{(k+1)}$. We then obtain the dominant eigenvalue by taking the limit

$$\lambda_1 = \lim_{k \to \infty} \frac{(y^{(k+1)})r}{(v^{(k)})r}$$

where λ represents the rth component of that vector. Obviously, there are n ratios of numbers. As $k \to \infty$ all these ratios tend to the same value, which is the largest eigenvalue in magnitude i.e., λ_1. The iteration is stopped when the magnitude of the difference of any two ratios is less than the prescribed tolerance.

The corresponding eigenvector is then $v^{(k+1)}$ obtained at the end of the last iteration performed.

We now illustrate the method through an example.

Example.2.15 Find the dominant eigenvalue and the corresponding eigenvector correct to two decimal places of the matrix.

$$A = \begin{bmatrix} 2 & -1 & 0 \\ -1 & 2 & -1 \\ 0 & -1 & 2 \end{bmatrix}$$

using the power method.

Solution We take

$$y^{(0)} = v^{(0)} = (1 \quad 1 \quad 1)^T$$

Using Equation, we obtain

$$y^{(1)} = Av^{(0)} = \begin{bmatrix} 2 & -1 & 0 \\ -1 & 2 & -1 \\ 0 & -1 & 2 \end{bmatrix}\begin{bmatrix} 1 \\ 1 \\ 1 \end{bmatrix} = \begin{bmatrix} 1 \\ 0 \\ 1 \end{bmatrix}$$

Now $m_1 = 1$ and $v^{(1)} = \frac{1}{m_1}y^{(1)} = (1 \quad 0 \quad 1)^T$

Again,

$$y^{(2)} = Av^{(1)} = \begin{bmatrix} 2 & -1 & 0 \\ -1 & 2 & -1 \\ 0 & -1 & 2 \end{bmatrix} \begin{bmatrix} 1 \\ 0 \\ 1 \end{bmatrix} = \begin{bmatrix} 2 \\ -2 \\ 2 \end{bmatrix}$$

$m_2 = 2$ and $v^{(2)} = \frac{1}{m_2} y^{(2)} = \frac{1}{2} y^{(2)} = (1 \quad -1 \quad 1)^T$.

Proceeding in this manner, we have

$$y^{(3)} = Av^{(2)} = [3 \quad -4 \quad 3]^T$$
$$m_3 = 4$$
$$v^{(3)} = \frac{1}{4} y^{(3)} = [0.75 \quad -1 \quad 0.75]^T$$
$$y^{(4)} = Av^{(3)} = [2.5 \quad -3.5 \quad 2.5]^T$$
$$m_4 = 3.5$$
$$v^{(4)} = \frac{1}{3.5} y^{(4)} = [0.7143 \quad -1 \quad 0.7143]^T$$
$$y^{(5)} = Av^{(4)} = [2.4286 \quad -3.4286 \quad 2.4286]^T$$
$$m_5 = 3.4286$$
$$v^{(5)} = \frac{1}{3.4286} y^{(5)} = [0.7083 \quad -1 \quad 0.7083]^T$$
$$y^{(6)} = Av^{(5)} = [2.4166 \quad -3.4166 \quad 2.4166]^T$$
$$m_6 = 3.4166$$
$$v^{(6)} = \frac{1}{3.4166} y^{(6)} = [0.7073 \quad -1 \quad 0.7073]^T$$
$$y^{(7)} = Av^{(6)} = [2.4146 \quad -3.4146 \quad 2.4146]^T$$
$$m_7 = 3.4146$$

$$v^{(7)} = \frac{1}{3.4146} y^{(7)} = [0.7071 \quad -1 \quad 0.7071]^T$$

After 7 iterations, the ratios $\frac{y^{(7)}r}{v^{(6)}r}$ are given as 3.4138, 3.4146, *and* 3.4138. The maximum error in these ratios is 0.0008. Hence the dominant eigenvalue can be taken as 3.414 and the corresponding eigenvector is $[0.7071 \quad -1 \quad 0.7071]^T$.

Note: that the exact dominant eigenvalue of A as obtained in Example 11.11 was $2 + \sqrt{2} = 3.4142$ and the corresponding Eigen vector was $[1 \quad -\sqrt{2} \quad 1]^T$ which can also be written as $[\frac{1}{\sqrt{2}} - 1 \frac{1}{\sqrt{2}}]^T = [0.7071 \quad -1 \quad 0.7071]^T$.

Interpolation

3.1. Introduction

Let $y = f(x), x_0 \leq x \leq x_n$ be defined. We can find the value of y for all values of x because $y = f(x)$ is defined explicitly. But without having the explicit definition of $y = f(x)$, it is difficult to find y.

If we have a set of tabular values

$$x = x_0, x_1, \ldots\ldots\ldots x_n$$
$$y = y_0, y_1, \ldots\ldots\ldots y_n$$

which satisfy $y = f(x)$, then we can find the value of y for the corresponding value of x, by using interpolation.

3.1.1. Interpolation

Interpolation is a method of constructing new data points from a discrete set of known data points.

i.e., Interpolation is the process of finding out the unknown value that lies in the given set of tabulated values.

Extrapolation is the process of finding out the unknown value which lies outside the given set of tabulated values.

3.1.2. Finite differences

Let $y = f(x)$ be a function in $x_0 \leq x \leq x_n$, x_i are equally spaced. Then we can recover the values of y for some intermediate values of x in the range $x_0 \leq x \leq x_n$ by using the 'difference' of $f(x)$.

The first finite difference of y is

$$\Delta y = \Delta f\left(x\right)$$
$$= f\left(x+\Delta x\right)-f\left(x\right)$$

where Δx is the increment in x.

$$\Delta^2 y = \Delta\left(\Delta y\right)=\Delta\left(f\left(x+\Delta x\right)-f\left(x\right)\right)$$
$$= f\left(x+2\Delta x\right)-2f\left(x+\Delta x\right)+f\left(x\right)$$

In general,

$$\Delta^n y = \Delta\left(\Delta^{n-1}y\right) \text{ for n=2, 3, 4, ...}$$

3.1.3. Forward difference

Let $y = f\left(x\right)$ be the continuous function. Let $y_0, y_1, y_2, \ldots y_n$ be the corresponding values of y at x respectively. Then the differences

$$y_1 - y_0, y_2 - y_1, \ldots y_n - y_{n-1},$$
i.e.,
$$\Delta y_0 = y_1 - y_0, \Delta y_1 = y_2 - y_1 \ldots \Delta y_{n-1} = y_n - y_{n-1},$$
$$i.e., \Delta\left(y_k\right) = y_{k+1} - y_k$$

Where Δ is called the forward difference operator.
The differences of the first forward differences are called second forward differences and are denoted by $\Delta^2 y_0, \Delta^2 y_1 \ldots\ldots$ similarly we can define higher order forward differences as follows.

$$\Delta^2 y_0 = \Delta y_1 - \Delta y_0 = y_2 - y_1 - \left(y_1 - y_0\right) = y_2 - 2y_1 + y_0$$
$$\Delta^2 y_1 = \Delta y_2 - \Delta y_1 = y_3 - y_2 - \left(y_2 - y_1\right) = y_3 - 2y_2 + y_1$$

The forward difference table is shown below

x	Y	Δ	Δ^2	Δ^3	Δ^4	Δ^5	Δ^6
x_0	y_0						
x_1	y_1	Δy_0					
x_2	y_2	Δy_1	$\Delta^2 y_0$				
x_3	y_3	Δy_2	$\Delta^2 y_1$	$\Delta^3 y_0$			
x_4	y_4	Δy_3	$\Delta^2 y_2$	$\Delta^3 y_1$	$\Delta^4 y_0$		
x_5	y_5	Δy_4	$\Delta^2 y_3$	$\Delta^3 y_2$	$\Delta^4 y_1$	$\Delta^5 y_0$	
x_6	y_6	Δy_5	$\Delta^2 y_4$	$\Delta^3 y_3$	$\Delta^4 y_2$	$\Delta^5 y_1$	$\Delta^6 y_0$

Δ can also be defined as $\Delta f(x) = f(x+h) - f(x)$

3.0.4. Backward difference

Let $y = f(x)$ be the continuous function. Let $y_0, y_1, y_2, \ldots y_n$ be the corresponding values of y at $x = x_0, x_1 \ldots x_n$ respectively. Then the differences $y_1 - y_0, y_2 - y_1, \ldots y_n - y_{n-1}$, are called the first backward differences. i.e.,

$\nabla y_0 = y_1 - y_0, \nabla y_1 = y_2 - y_1 \ldots \nabla y_{n-1} = y_n - y_{n-1}$,

$i.e., \nabla(y_k) = y_k - y_{k-1}$

Where ∇ is called the backward difference operator.
The differences of the first backward differences are called second backward differences and they are denoted by $\nabla^2 y_2, \nabla^2 y_3 \ldots \ldots \nabla^2 y_n$. similarly, we can define higher order backward differences as follows.

$$\nabla^2 y_2 = \nabla y_2 - \nabla y_1 = y_2 - y_1 - (y_1 - y_0) = y_2 - 2y_1 + y_0$$

$$\nabla^3 y_3 = \nabla^2 y_3 - \nabla^2 y_2 = y_3 - 3y_2 + 3y_1 - y_0$$

and so on.

The Backward difference table is shown below

x	Y	∇	∇^2	∇^3	∇^4	∇^5	∇^6
x_0	y_0						
x_1	y_1	∇y_1					
x_2	y_2	∇y_2	$\nabla^2 y_2$				
x_3	y_3	∇y_3	$\nabla^2 y_3$	$\nabla^3 y_3$			
x_4	y_4	∇y_4	$\nabla^2 y_4$	$\nabla^3 y_4$	$\nabla^4 y_4$		
x_5	y_5	∇y_5	$\nabla^2 y_5$	$\nabla^3 y_5$	$\nabla^4 y_5$	$\nabla^5 y_5$	
x_6	y_6	∇y_6	$\nabla^2 y_6$	$\nabla^3 y_6$	$\nabla^4 y_6$	$\nabla^5 y_6$	$\nabla^6 y_6$

∇ can also be defined as $\nabla f(x) = f(x) - f(x-h)$

3.2. Newton's forward and difference interpolation formula

Let $y = f(x)$ be the continuous function. Let $y_0, y_1, y_2, \ldots y_n$ be the corresponding values of y at $x = x_0, x_1 \ldots x_n$ respectively. Here x values are equally spaced with common difference h. Consider $f(x)$ as an nth degree polynomial in x.

Let

$$y = A_0 + A_1(x - x_0) + A_2(x - x_0)(x - x_1)$$
$$+ A_n(x - x_0)(x - x_1)(x - x_2)\ldots\ldots(x - x_{n-1})\ldots(3.1)$$

Where $A_0, A_1, A_2, \ldots A_n$ are constants.

Put $x = x_0$ in (3.1)

$$y = f(x_0) = A_0 \Rightarrow y_0 = A_0$$

Put $x = x_1$ in (3.1)

$$y = f(x_1) = A_0 + A_1(x_1 - x_0) \Rightarrow y_1 = y_0 + A_1 h$$

$$\Rightarrow A_1 h = y_1 - y_0 \Rightarrow A_1 = \frac{\Delta y_0}{h}$$

Put $x = x_2$ in (3.1)

$$y = f(x_2) = A_0 + A_1(x_2 - x_0) + A_2(x_2 - x_0)(x_2 - x_1)$$

$$\Rightarrow y_2 = y_0 + 2A_2 h^2$$

$$\Rightarrow 2A_2 h^2 = y_2 - y_0 - 2\Delta y_0 \Rightarrow A_2 = \frac{\Delta^2 y_0}{\angle 2 h^2}$$

Similarly, $A_3 = \dfrac{\Delta^3 y_0}{\angle 3 h^3} \ldots\ldots\ldots\ldots\ldots A_n = \dfrac{\Delta^n y_0}{\angle n h^n}$

Put these values in (3.1)

$$y = y_0 + \frac{\Delta y_0}{h}(x - x_0) + \frac{\Delta^2 y_0}{\angle 2 h^2}(x - x_0)(x - x_1) + \ldots\ldots$$

$$+ \frac{\Delta^n y_0}{\angle n h^n}(x - x_0)(x - x_1)\ldots(x - x_{n-1})\ldots(3.2)$$

Put $x = x_0 + ph$ where $p = \dfrac{x - x_0}{h}$ in (3.2)

$$y = f(x_0 + ph) = y_0 + (x_0 + ph - x_0)\frac{\Delta y_0}{h}$$
$$+ (x_0 + ph - x_0)(x_0 + ph - x_1)$$
$$\frac{\Delta^2 y_0}{\angle 2h^2} + \dots + (x_0 + ph - x_0)(x_0 + ph - x_1)$$
$$(x_0 + ph - x_{n-1})\frac{\Delta^n y_0}{\angle n h^n}$$

$$\Rightarrow y = y_0 + ph\frac{\Delta y_0}{h} + ph(ph - h)\frac{\Delta^2 y_0}{\angle 2h^2} + \dots +$$
$$ph(ph - h)(ph - 2h)\dots(ph - (n-1)h)\frac{\Delta^n y_0}{\angle n h^n}$$
$$\left[\because x_1 - x_0 = h\right]$$

$$\Rightarrow y = y_0 + p\Delta y_0 + p(p-1)\frac{\Delta^2 y_0}{\angle 2} + \dots +$$
$$p(p-1)(p-2)\dots(p - (n-1))\frac{\Delta^n y_0}{\angle n}$$

The above formula is called Newton's forward formula difference interpolation formula.

Note.1. It is used to interpolate the values of y near the beginning of a set of tabular values.

1. y_0 may be taken as any point of the table, this formula is used to find y which come after the value chosen as y_0.

3.3. Newton's backward and difference interpolation formula

Let $y = f(x)$ be the continuous function. Let $y_0, y_1, y_2, \ldots y_n$ be the corresponding values of y at $x = x_0, x_1 \ldots x_n$ respectively. Here x values are equally spaced with common difference h. Consider $f(x)$ as an nth degree polynomial in x. Let

$$y = A_0 + A_1(x - x_n) + A_2(x - x_n)(x - x_{n-1}) + \ldots +$$
$$A_n(x - x_n)(x - x_{n-1})(x - x_{n-2}) \ldots (x - x_1) \ldots (3.3)$$

Where $A_0, A_1, A_2, \ldots A_n$ are constants to be determined.

Put $x = x_n$ in (3.3)

$$y = f(x_n) = A_0 \Rightarrow y_n = A_0$$

Put $x = x_{n-1}$ in (3.3)

$$y = f(x_{n-1}) = A_0 + A_1(x_{n-1} - x_n) \Rightarrow y_{n-1} = y_n + A_1 h$$

$$\Rightarrow -A_1 h = y_{n-1} - y_n \Rightarrow A_1 = \frac{\nabla y_n}{h}$$

Similarly,

$$A_2 = \frac{\nabla^2 y_n}{\angle 2 h^2}, A_3 = \frac{\nabla^3 y_0}{\angle 3 h^3} \ldots \ldots \ldots \ldots \ldots A_n = \frac{\nabla^n y_0}{\angle n h^n}$$

Put these values in (3.3)

$$y = y_n + \frac{\nabla y_n}{h}(x - x_n) + \frac{\nabla^2 y_n}{\angle 2 h^2}(x - x_n)(x - x_{n-1}) + \ldots \ldots$$

$$+ \frac{\nabla^n y_0}{\angle n h^n}(x - x_n)(x - x_{n-1}) \ldots (x - x_1) \ldots$$

104

Put $x = x_n + ph$ where $p = \dfrac{x - x_n}{h}$ in the above equation

$$y = f(x_n + ph) = y_n + (x_n + ph - x_n)\dfrac{\nabla y_n}{h}$$

$$+ (x_n + ph - x_n)(x_n + ph - x_{n-1})\dfrac{\nabla^2 y_n}{\angle 2h^2}$$

$$+ .... + (x_n + ph - x_n)(x_n + ph - x_{n-1})(x_n + ph - x_1)\dfrac{\nabla^n y_n}{\angle n h^n}$$

$$\Rightarrow y = y_n + p\nabla y_n + p(p+1)\dfrac{\nabla^2 y_n}{\angle 2} + ...$$

$$+ p(p+1)(p+2)....(p-(n-1))\dfrac{\nabla^n y_n}{\angle n}$$

where $p = \dfrac{x - x_n}{h}$

The above formula is called Newton's backward formula difference interpolation formula. It is used for interpolating a value of y nearer to the end of the table of values.

Example.3.1 Find Solution using Newton's Forward Difference formula

x	f(x)
1891	46
1901	66
1911	81

1921	93
1931	101

x   =   1895,   finding   the   Value   f(2).

Solution

Given

x	1891	1901	1911	1921	1931
y	46	66	81	93	101

Newton's forward difference interpolation method to find solution. Then Newton's forward difference table is

x	y	Δy	$\Delta 2y$	$\Delta 3y$	$\Delta 4y$
1891	46				
		66-46=20			
1901	66		15-20=-5		
		81-66=15		-3--5=2	
1911	81		12-15=-3		-1-2=-3
		93-81=12		-4--3=-1	

106

1921	93		8-12=-4		
		101-93=8			
1931	101				

The value of x at you want to find the $f(x):x=1$

$$h = x_1 - x_0 = 1901 - 1891 = 10$$

$$p = \frac{x - x_0}{h} = \frac{1895 - 1891}{10} = 0.4$$

Newton's forward difference interpolation formula is

$$\Rightarrow y = y_0 + p\Delta y_0 + p(p-1)\frac{\Delta^2 y_0}{\angle 2} + ... + p(p-1)(p-2)....(p-(n-1))\frac{\Delta^n y_0}{\angle n}$$

$$y(1895)=46+8+0.6+0.128+0.1248$$

$$y(1895)=54.8528$$

Solution of newton's forward interpolation method $y(1895)=54.8528$

Example.3.2 Find Solution using Newton's Forward Difference formula

x	f(x)
0	1
1	0
2	1
3	10

find at x = -1

Solution

Given

x	0	1	2	3
y	1	0	1	10

Newton's forward difference interpolation method to find solution. Newton's forward difference table is

x	y	Δy	$\Delta 2y$	$\Delta 3y$
0	1			
		-1		

1	0		2	
		1		6
2	1		8	
		9		
3	10			

$$h = x_1 - x_0 = 1 - 0 = 1$$

$$p = \frac{x - x_0}{h} = \frac{-1-0}{1} = -1$$

Newton's forward difference interpolation formula is

$$\Rightarrow y = y_0 + p\Delta y_0 + p(p-1)\frac{\Delta^2 y_0}{\angle 2} + ... + p(p-1)(p-2)....(p-(n-1))\frac{\Delta^n y_0}{\angle n}$$

y(-1)=1+(-1)×-1+-1(-1-1)2×2+-1(-1-1)(-1-2)6×6

y(-1)=1+1+2-6

y(-1)=-2

Example.3.3 Find Solution of an equation $x^3 - x + 1 = 0$ using Newton's Forward Difference formula $x_1 = 2$ and $x_2 = 4$ find x = 2.25, Step value (h) = 0.5, finding f(2).

Solution Let $f(x) = x^3 - x + 1$

x	2	2.5	3	3.5	4
y	7	14.125	25	40.375	61

Newton's forward difference interpolation method to find solution

Newton's forward difference table is

x	y	Δy	$\Delta 2y$	$\Delta 3y$	$\Delta 4y$
2	7				
		7.125			
2.5	14.125		3.75		
		10.875		0.75	
3	25		4.5		0
		15.375		0.75	
3.5	40.375		5.25		
		20.625			
4	61				

The value of x at you want to find the $f(x):x=2.25$

$$h = x_1 - x_0 = 2.5 - 2 = 0.5$$

$$p = \frac{x - x_0}{h} = \frac{2.25 - 20.5}{0.5} = 0.5$$

Newton's forward difference interpolation formula is

$$\Rightarrow y = y_0 + p\Delta y_0 + p(p-1)\frac{\Delta^2 y_0}{\angle 2} + ... + p(p-1)(p-2)....$$

$$(p-(n-1))\frac{\Delta^n y_0}{\angle n}$$

$$y(2.25) = 7 + 0.5 \times 7.125 + 0.5(0.5-1)2 \times 3.75$$

$$+0.5(0.5-1)(0.5-2)6 \times 0.75 + 0.5(0.5-1)(0.5-2)(0.5-3)24 \times 0$$

$$y(2.25) = 10.1406$$

Solution of newton's forward interpolation method $y(2.25)=10.1406$

Example.3.4. Find Solution of an equation $2x^3 - 4x + 1$ using Newton's Forward Difference formula $x_1 = 2$ and $x_2 = 4$, find x = 2.1, Step value (h) = 0.25. Finding f(2).

Solution Let $f(x) = 2x^3 - 4x + 1$

x	2	2.25	2.5	2.75	3	3.25	3.5	3.75	4
y	9	14.7812	22.25	31.5938	43	56.6562	72.75	91.4688	113

Newton's forward difference interpolation method to find solution. Newton's forward difference table is

111

x	y	Δy	$\Delta 2y$	$\Delta 3y$	$\Delta 4y$
2	9				
		5.7812			
2.25	14.7812		1.6875		
		7.4688		0.1875	
2.5	22.25		1.875		0
		9.3438		0.1875	
2.75	31.5938		2.0625		0
		11.4062		0.1875	
3	43		2.25		0
		13.6562		0.1875	
3.25	56.6562		2.4375		0
		16.0938		0.1875	
3.5	72.75		2.625		0
		18.7188		0.1875	

3.75	91.4688		2.8125		
		21.5312			
4	113				

The value of x at you want to find the $f(x)$: $x=2.1$

$$h = x_1 - x_0 = 2.25 - 2 = 0.25$$

$$p = \frac{x - x_0}{h} = \frac{2.1 - 20.2}{5} = 0.4$$

Newton's forward difference interpolation formula is

$$\Rightarrow y = y_0 + p\Delta y_0 + p(p-1)\frac{\Delta^2 y_0}{\angle 2} + \dots + p(p-1)(p-2)\dots$$

$$(p-(n-1))\frac{\Delta^n y_0}{\angle n}$$

$y(2.1)=9+0.4\times5.7812+0.4(0.4-1)2\times1.6875+0.4(0.4-1)(0.4-2)6\times0.1875+0.4(0.4-1)(0.4-2)(0.4-3)24\times0$

$y(2.1)=9+2.3125-0.2025+0.012+0$

$y(2.1)=11.122$

Solution of newton's forward interpolation method $y(2.1)=11.122$

Example.3.5

The values of sinx are given below for different values of x. Find sin32.

X	30	35	40	45	50
Y	0.5	0.5736	0.6428	0.7071	0.7600

Solution The value of sin32 is nearer to the beginning of the table.Therefore, we use Newton's forward difference interpolation formula.

x	Y	Δ	Δ^2	Δ^3	Δ^4
30	0.5				
35	0.5736	0.0736			
40	0.6428	0.0692	-0.0044		
45	0.7071	0.0643	-0.0049	-0.0005	
50	0.7660	0.0589	-0.0054	-0.0005	0

By Newton's forward formula,

$$y = y_0 + p\Delta y_0 + p(p-1)\frac{\Delta^2 y_0}{\angle 2} + ...$$
$$+ p(p-1)(p-2)....(p-(n-1))\frac{\Delta^n y_0}{\angle n},$$

n=1, 2, 3,.....

Here $x_0 = 30^0, h = 5, x = 32^0$,

$$\Rightarrow p = \frac{x - x_0}{h} = \frac{32 - 30}{5} = 0.4$$

$$\Rightarrow y = 0.5 + (0.4)(0.0736) + \frac{0.4(0.4-1)}{2}(-0.0044)$$

$$+0.4\frac{(0.4-1)(0.4-2)}{6}(-0.0005) + 0$$

$$\Rightarrow y = 0.529936$$

Example.3.6 The population of a town in decimal census was as given below. Estimate the population for the year 1895.

Year x	1891	1901	1911	1921	1931
Population y in thousands	46	66	81	93	101

Solution The difference table is given below

x	Y	Δ	Δ^2	Δ^3	Δ^4
1891	46				
1901	66	20			
1911	81	15	-5		
1921	93	12	-3	-2	
1931	101	8	-4	-1	-3

Here $x_0 = 1891, h = 10, x = 1895, p = \dfrac{x - x_0}{h} = 0.4$

Newton's forward formula, for n=1,2,3.... is

$$y = y_0 + p\Delta y_0 + p(p-1)\frac{\Delta^2 y_0}{\angle 2} + ... + p(p-1)(p-2)....$$

$$\left(p - (n-1)\right)\frac{\Delta^n y_0}{\angle n}$$

$$\Rightarrow y = 46 + 0.4(20) + \frac{0.4(0.4-1)}{2}(-5)$$

$$+\frac{0.4(0.4-1)(0.4-2)}{6}(2) + \frac{0.4(0.4-1)(0.4-2)(0.4-3)}{24}(-3)$$

$$y = 48 + 8 + 0.6 + 0.128 + 0.1248$$

$$y = 54.8528$$

Example.3.7 Find the cubic polynomial which takes the following values.

$$y(0)=1, y(1)=0, y(2)=1, y(3)=10.$$

Solution

x	Y	Δ	Δ^2	Δ^3
0	1			
1	0	-1		
2	1	1	2	
3	10	9	8	6

Here $\quad x_0 = 0, h = 1, p = \dfrac{x-x_0}{h} = x$

Newton's forward formula is

$$y = y_0 + p\Delta y_0 + p(p-1)\frac{\Delta^2 y_0}{\angle 2} + ... + p(p-1)(p-2)....$$

$$(p-(n-1))\frac{\Delta^n y_0}{\angle n}$$

,n=1,2,3….

$$\Rightarrow y = 1 + x(-1) + \frac{x(x-1)}{2}.2 + \frac{x(x-1)(x-2)}{6}.6$$

$$\Rightarrow y = x^3 - 2x^2 + 1$$

Note. There may be polynomials of higher degree which also fit the data, but Newton's formula gives us the polynomial of least degree which fits the data.

Example.3.8 Calculate the value $f(7.5)$ for the table

X	1	2	3	4	5	6	7	8
Y	1	8	27	64	125	216	343	512

Solution Since the value of x is nearer to the end of the table, we use the Newton's backward difference formula,

x	Y	∇	∇^2	∇^3	∇^4	∇^5	∇^6
1	1						
2	8	7					
3	27	19	12				
4	64	37	18	6			
5	125	61	24	6	0		
6	216	91	30	6	0	0	
7	343	127	36	6	0	0	0
8	512	169	42	6	0	0	0

Here $x_n = 8, h = 1, x = 7.5, p = \dfrac{x - x_n}{h} = -0.5$

Newton's   backward   difference   formula

$$\Rightarrow y = y_n + p\nabla y_n + p(p+1)\frac{\nabla^2 y_n}{\angle 2} + ... + p(p+1)(p+2)....(p-(n-1))\frac{\nabla^n y_n}{\angle n}$$

$$\Rightarrow y = 512 + (-0.5)169 + \frac{(-0.5)(-0.5+1)}{2}42$$

$$+ \frac{(-0.5)(-0.5+1)(-0.5+2)}{3}6$$

$$\Rightarrow y = 512 - 84.5 - 5.25 - 0.375$$

$$\Rightarrow y = 421.875$$

Example.3.9 The area A of a circle of diameter d is given below.

117

D	80	85	90	95	100
A	5026	5674	6362	7088	7854

Find approximately the areas of circles of diameters 82 and 91.

Solution The difference table is given below

d	a	Δ	Δ^2	Δ^3	Δ^4
80	5026				
85	5674	648			
90	6362	688	40		
95	7088	726	38	-2	
100	7854	766	40	2	4

Here $x_0 = 80, h = 5, x = 82, p = \dfrac{x - x_0}{h} = \dfrac{2}{5} = 0.4$

By using Newton's forward formula, for n=1,2,3,...

$$\Rightarrow y = 5026 + 0.4(648) + \frac{0.4(0.4-1)40}{2}$$

$$+ \frac{0.4(0.4-1)(0.4-2)(-2)}{6}$$

$$\Rightarrow y = 5280.1056$$

By using Newton's Backward formula,

$$\Rightarrow y = y_n + p\nabla y_n + p(p+1)\frac{\nabla^2 y_n}{\angle 2} + ... + p(p+1)(p+2)....(p-(n-1))\frac{\nabla^n y_n}{\angle n}$$

$$\Rightarrow y = 7854 + (-1.8)(766) + \frac{(-1.8)(-1.8+1)}{2}40$$

$$+ \frac{(-1.8)(-1.8+1)(-1.8+2)}{6}2$$

$$+ \frac{(-1.8)(-1.8+1)(-1.8+2)(-1.8+3)4}{24}$$

$$\Rightarrow y = 6504.1536.$$

Example.3.10 Find Solution using Newton's Backward Difference formula find x=1925.

x	f(x)
1891	46
1901	66
1911	81
1921	93
1931	101

Solution

x	1891	1901	1911	1921	1931
y	46	66	81	93	101

Newton's backward difference interpolation method to find solution. Newton's backward difference table is

x	y	∇y	$\nabla 2y$	$\nabla 3y$	$\nabla 4y$
1891	46				
		20			

1901	66		-5	
		15		2
1911	81		-3	**-3**
		12	**-1**	
1921	93		**-4**	
		8		
1931	**101**			

The value of x at you want to find the $f(x):x=1925$

$$h = x_1 - x_0 = 1901 - 1891 = 10$$

$$p = \frac{x - x_n}{h} = \frac{1925 - 1931}{10} = -0.6$$

Newton's backward difference interpolation formula is

$$\Rightarrow y = y_n + p\nabla y_n + p(p+1)\frac{\nabla^2 y_n}{\angle 2} + \dots + p(p+1)(p+2)\dots$$

$$(p-(n-1))\frac{\nabla^n y_n}{\angle n}$$

$$y(1925) = 101 - 4.8 + 0.48 + 0.056 + 0.1008$$

$$y(1925) = 96.8368$$

120

Solution of newton's backward interpolation method $y(1925)=96.8368$

Example.3.10 Find Solution using Newton's Backward Difference formula. Find f(4).

x	f(x)
0	1
1	0
2	1
3	10

Solution Let

x	0	1	2	3
y	1	0	1	10

Newton's backward difference interpolation method to find solution. Newton's backward difference table is

x	y	∇y	$\nabla 2y$	$\nabla 3y$
0	1			
		-1		
1	0		2	
		1		6
2	1		8	
		9		
3	10			

The value of x at you want to find the $f(x)$: $x=4$

$$h = x_1 - x_0 = 1 - 0 = 1$$

$$p = \frac{x - x_n}{h} = \frac{4-3}{1} = 1$$

Newton's backward difference interpolation formula is

$$\Rightarrow y = y_n + p\nabla y_n + p(p+1)\frac{\nabla^2 y_n}{\angle 2} + ...$$

$$+ p(p+1)(p+2)....(p-(n-1))\frac{\nabla^n y_n}{\angle n}$$

$y(4)=10+1\times 9+1(1+1)2\times 8+1(1+1)(1+2)6\times 6$

$y(4)=10+9+8+6$

122

$y(4)=33$

Solution of newton's backward interpolation method $y(4)=33$

Example.3.11 Find Solution of an equation $x^3 - x + 1$ using Newton's Backward Difference formula $x_1 = 2$ and $x_2 = 4$, $x = 3.75$, Step value (h) = 0.5, finding f(2).

Solution Given equation $f(x) = x^3 - x + 1$. Let

x	2	2.5	3	3.5	4
y	7	14.125	25	40.375	61

Newton's backward difference interpolation method to find solution Newton's backward difference table is

x	y	∇y	$\nabla^2 y$	$\nabla^3 y$	$\nabla^4 y$
2	7				
		7.125			
2.5	14.125		3.75		
		10.875		0.75	
3	25		4.5		0
		15.375		0.75	

123

3.5	40.375		5.25		
		20.625			
4	61				

The value of x at you want to find the $f(x)$: $x=3.75$

$$h = x_1 - x_0 = 2.5 - 2 = 0.5$$

$$p = \frac{x - x_n}{h} = \frac{3.75 - 40.5}{1} = -0.5$$

Newton's backward difference interpolation formula is

$$y = y_n + p\nabla y_n + p(p+1)\frac{\nabla^2 y_n}{\angle 2} + \dots$$

$$+ p(p+1)(p+2)\dots(p-(n-1))\frac{\nabla^n y_n}{\angle n}$$

$$y(3.75) = 61 + (-0.5) \times 20.625 - 0.5(-0.5+1)2$$
$$\times 5.25 + 0.5(-0.5+1)(-0.5+2)6 \times 0.75$$
$$- 0.5(-0.5+1)(-0.5+2)(-0.5+3)24 \times 0$$

$y(3.75) = 61 - 10.3125 - 0.6562 - 0.0469 + 0$

$y(3.75) = 49.9844$

Solution of newton's backward interpolation method $y(3.75) = 49.9844$

Example.3.12 Find Solution of an equation $2x^3 - 4x + 1$ using Newton's Backward Difference formula $x_1 = 2$ and $x_2 = 4$, $x = 3.75$, step value (h) = 0.5, finding f(2).

Solution: Given $f(x) = 2x^3 - 4x + 1$. Then

124

x	2	2.5	3	3.5	4
y	9	22.25	43	72.75	113

Newton's backward difference interpolation method to find solution. Newton's backward difference table is

x	y	∇y	$\nabla^2 y$	$\nabla^3 y$	$\nabla^4 y$
2	9				
		13.25			
2.5	22.25		7.5		
		20.75		1.5	
3	43		9		**0**
		29.75		**1.5**	
3.5	72.75		**10.5**		
		40.25			
4	113				

The value of x at you want to find the $f(x):x=3.75$

$$h = x_1 - x_0 = 2.5 - 2 = 0.5$$

$$p = \frac{x - x_n}{h} = \frac{3.75 - 40.5}{1} = -0.5$$

Newton's backward difference interpolation formula is

$$\Rightarrow y = y_n + p\nabla y_n + p(p+1)\frac{\nabla^2 y_n}{\angle 2} + \ldots$$

$$+ p(p+1)(p+2)\ldots(p-(n-1))\frac{\nabla^n y_n}{\angle n}$$

$$y(3.75) = 113 + (-0.5) \times 40.25 + -0.5(-0.5+1)2 \times 10.5$$
$$\qquad + -0.5(-0.5+1)$$
$$(-0.5+2)6 \times 1.5 + -0.5(-0.5+1)(-0.5+2)(-0.5+3)24 \times 0$$
$$y(3.75) = 113 - 20.125 - 1.3125 - 0.0938 + 0$$
$$y(3.75) = 91.4688$$

Solution of newton's backward interpolation method

$y(3.75) = 91.4688$

3.4. Central difference interpolation formula

Up to now, we have discussed Newton's forward and backward interpolation formula which are applicable to find the values nearer to the beginning and end of the table respectively. Now we discuss the central difference formula which are most suited for interpolation nearer to the middle of the tabulated set.

Interpolation with unevenly spaced points.

In the proceeding sections, we have derived various interpolation formula and discussed their uses practically. The previous formulae possess a disadvantage that they require the values of the independent variable to be equally spaced. In some cases, it is desirable to have interpolation formula with unequally spaced values of the argument. Here we introduce Lagrange's interpolation formula which uses only the function values,

126

3.5. Lagrange's interpolation formula

Let $y = f(x)$ be a function. Let $f(x_0), f(x_1), f(x_2), \dots f(x_n)$ be the values corresponding to $x = x_0, x_1, x_2, \dots x_n$, where the values of x are not equally spaced. Since n+1 values of $f(x)$ are given for n+1 values of x, we can take $y = f(x)$ as a polynomial in x of degree n.

$$y = A_0 (x-x_1)(x-x_2)\dots(x-x_n) + A_1 (x-x_0)(x-x_2)\dots(x-x_n)$$
$$+ \dots + A_n (x-x_0)(x-x_1)(x-x_2)\dots\dots(x-x_{n-1})$$

Put $x = x_0$ in (3.5)

Then $A_0 = \dfrac{f(x_0)}{(x_0 - x_1)(x_0 - x_2)\dots(x_0 - x_n)}$

Put $x = x_1$ in (3.5)

Then $A_1 = \dfrac{f(x_1)}{(x_1 - x_0)(x_1 - x_2)\dots(x_1 - x_n)}$

Similarly $A_2 = \dfrac{f(x_2)}{(x_2 - x_0)(x_2 - x_1)\dots(x_2 - x_n)}$,

$A_3 = \dfrac{f(x_3)}{(x_3 - x_0)(x_3 - x_1)\dots(x_3 - x_n)}$ ....

$A_n = \dfrac{f(x_n)}{(x_n - x_0)(x_n - x_1)\dots(x_n - x_{n-1})}$

Put all these values in (3.5)

$$y = f(x) = \frac{(x-x_1)(x-x_2)....(x-x_n)}{(x_0-x_1)(x_0-x_2)....(x_0-x_n)} f(x_0)$$

$$+ \frac{(x-x_0)(x-x_2)....(x-x_n)}{(x_1-x_0)(x_1-x_2)....(x_1-x_n)} f(x_1)$$

$$+............\frac{(x-x_0)(x-x_1)....(x-x_{n-1})}{(x_n-x_0)(x_n-x_1)....(x_n-x_{n-1})} f(x_n)$$

This is called Lagrange's interpolation formula. It is simple and easy to remember but the application of the formula is not speedy and there is a chance of committing mistakes due to number of positive and negative terms in numerator and denominator. If the x terms are equally spaced, then Lagrange's interpolation polynomial will coincide with the Newton's interpolation polynomial.

Example.3.13 Using Lagrange's interpolation formula, find y corresponding to x=10.

X	5	6	9	11
Y	12	13	14	16

Solution Given that

$$x_0 = 5, x_1 = 6, x_2 = 9, x_3 = 11$$
$$f(x_0) = 12, f(x_1) = 13, f(x_2) = 14, f(x_3) = 16$$

By lagrange's interpolation formula

128

$$y = f(x) = \frac{(x-x_1)(x-x_2)....(x-x_n)}{(x_0-x_1)(x_0-x_2)....(x_0-x_n)} f(x_0)$$

$$+\frac{(x-x_0)(x-x_2)....(x-x_n)}{(x_1-x_0)(x_1-x_2)....(x_1-x_n)} f(x_1)$$

$$+............\frac{(x-x_0)(x-x_1)....(x-x_{n-1})}{(x_n-x_0)(x_n-x_1)....(x_n-x_{n-1})} f(x_n)$$

n=1,2,3...

$$y = \frac{(4)(1)(-1)}{(-1)(-4)(-6)}12 + \frac{5.1.(-1)}{1.-3.-5}13 + \frac{5.4.-1}{4.3.-2}14 + \frac{5.4.1}{6.5.2}16$$

$$= 14.6666$$

Example.3.13 Certain values of x and $\log_{10}^x$ are

$(300, 2.4771)(304, 2.4829)(305, 2.4843)(307, 2.4871)$.

Find $\log_{10}^{301}$.

Solution Given that

$x_0 = 300, x_1 = 304, x_2 = 305, x_3 = 307, x = 301$

$f(x_0) = 2.4771, f(x_1) = 2.4829,$
$f(x_2) = 2.4843, f(x_3) = 2.4871$

By lagrange's interpolation formula

$$y = f(x) = \frac{(x - x_1)(x - x_2)....(x - x_n)}{(x_0 - x_1)(x_0 - x_2)....(x_0 - x_n)} f(x_0)$$

$$+ \frac{(x - x_0)(x - x_2)....(x - x_n)}{(x_1 - x_0)(x_1 - x_2)....(x_1 - x_n)} f(x_1)$$

$$+ .... \frac{(x - x_0)(x - x_1)....(x - x_{n-1})}{(x_n - x_0)(x_n - x_1)....(x_n - x_{n-1})} f(x_n)$$

n=1,2,3…

$$y = \frac{(-3)(-4)(-6)}{(-4)(-5)(-7)} 2.4771 + \frac{1. - 4.(-6)}{4. - 1. - 3} 2.4829$$

$$+ \frac{1. - 3. - 6}{5.1. - 2} 2.4843 + \frac{1. - 3. - 4}{7.3.2} 2.4871$$

$$= 2.47856$$

Example.3.14 Given the tables of values

x	150	152	154	156
$y = \sqrt{x}$	12.247	12.329	12.410	12.49

Solution Given

$x_0 = 150, x_1 = 152, x_2 = 154, x_3 = 156, x = 155$

$f(x_0) = 12.247, f(x_1) = 12.329, f(x_2) = 12.410, f(x_3) = 12.49$

By lagrange's interpolation formula

$$y = f(x) = \frac{(x-x_1)(x-x_2)....(x-x_n)}{(x_0-x_1)(x_0-x_2)....(x_0-x_n)} f(x_0)$$

$$+ \frac{(x-x_0)(x-x_2)....(x-x_n)}{(x_1-x_0)(x_1-x_2)....(x_1-x_n)} f(x_1)$$

$$+............\frac{(x-x_0)(x-x_1)....(x-x_{n-1})}{(x_n-x_0)(x_n-x_1)....(x_n-x_{n-1})} f(x_n)$$

n=1,2,3...

$$y = \frac{3.1.-1}{-2.-4.-6} 12.247 + \frac{5.1.-1}{2.-2.-4} 12.329$$

$$+ \frac{5.3.-1}{4.2.-2} 12.410 + \frac{5.3.1}{6.4.2} 12.49$$

$$\therefore y = 12.45$$

Example.3.15 Using Lagrange's interpolation formula, express

$$\frac{3x^2+x+1}{(x-1)(x-2)(x-3)}$$ as sum of partial fractions.

Solution Let $f(x) = 3x^2 + x + 1$. Then

$$(x-1)(x-2)(x-3) = 0 \Rightarrow x = 1,2,3$$
$$x_0 = 1, x_1 = 2, x_2 = 3$$
$$f(x_0) = 5, f(x_1) = 15, f(x_2) = 31$$

By lagrange's interpolation formula

$$y = f(x) = \frac{(x-x_1)(x-x_2)....(x-x_n)}{(x_0-x_1)(x_0-x_2)....(x_0-x_n)} f(x_0)$$

$$+ \frac{(x-x_0)(x-x_2)....(x-x_n)}{(x_1-x_0)(x_1-x_2)....(x_1-x_n)} f(x_1)$$

$$+............ \frac{(x-x_0)(x-x_1)....(x-x_{n-1})}{(x_n-x_0)(x_n-x_1)....(x_n-x_{n-1})} f(x_n)$$

n=1,2,3...

$$3x^2 + x + 1 = \frac{(x-2)(x-3)}{-1.-2} 5 + \frac{(x-1)(x-3)}{1.-1} 15$$

$$+ \frac{(x-1)(x-2)}{2.1} 31$$

Divide with $(x-1)(x-2)(x-3)$

$$\frac{3x^2 + x + 1}{(x-1)(x-2)(x-3)} = \frac{5}{2(x-1)} - \frac{15}{x-2} + \frac{31}{2(x-3)}$$

Using Lagrange's interpolation formula find y(10) from the following table:

X	5	6	9	11
Y	12	13	14	16

Here the intervals are unequal. By Lagrange's interpolation formula we have

$$x_0 = 5, x_1 = 6, x_2 = 9, x_3 = 11$$

$$y_0 = 12, y_1 = 13, y_2 = 14, y_3 = 16$$

$$y = f(x) = \frac{(x-x_1)(x-x_2)(x-x_3)}{(x_0-x_1)(x_0-x_2)(x_0-x_3)} \times y_0 + \frac{(x-x_0)(x-x_2)(x-x_3)}{(x_1-x_0)(x_1-x_2)(x_1-x_3)} \times y_1$$

$$+ \frac{(x-x_0)(x-x_1)(x-x_3)}{(x_2-x_0)(x_2-x_1)(x_2-x_3)} \times y_2 + \frac{(x-x_0)(x-x_1)(x-x_2)}{(x_3-x_0)(x_3-x_1)(x_3-x_2)} \times y_3$$

$$= \frac{(x-6)(x-9)(x-11)}{(5-6)(5-6)(5-11)}(12) + \frac{(x-5)(x-9)(x-11)}{(6-5)(6-9)(6-9)}(13)$$

$$+ \frac{(x-5)(x-6)(x-11)}{(9-5)(9-6)(9-11)}(14) + \frac{(x-5)(x-6)(x-9)}{(11-5)(11-6)(11-9)}(16)$$

Put $x = 10$

$$y(10) = f(10) = \frac{4(1)(-1)}{(-1)(-4)(-6)}(12) + \frac{(5)(1)(-1)}{(1)(-3)(-5)}(13) + \frac{5(4)(-1)}{4(3)(-2)}(14) + \frac{(5)(4)(1)}{6(5)(2)}(16)$$

$$= \frac{1}{6}(12) - \frac{13}{3} + \frac{5(14)}{3\times2} + \frac{4\times16}{12}$$

$$= 14.6663$$

Example.3.16 Find Solution using Lagrange's Interpolation formula

x	f(x)
300	2.4771
304	2.4829
305	2.4843
307	2.4871

find x=301.

133

Solution Given

x	300	304	305	307
y	2.4771	2.4829	2.4843	2.4871

Lagrange's interpolating polynomial

The value of x at you want to find $Pn(x)$:$x=301$

$$y = f(x) = \frac{(x-x_1)(x-x_2)....(x-x_n)}{(x_0-x_1)(x_0-x_2)....(x_0-x_n)} f(x_0)$$

$$+ \frac{(x-x_0)(x-x_2)....(x-x_n)}{(x_1-x_0)(x_1-x_2)....(x_1-x_n)} f(x_1)$$

$$+............\frac{(x-x_0)(x-x_1)....(x-x_{n-1})}{(x_n-x_0)(x_n-x_1)....(x_n-x_{n-1})} f(x_n)$$

Lagrange's formula is

$y(301)$ = (301-304)(301-305)(301-307)(300-304)(300-305)

(300-307)×2.4771+(301-300)(301-305)(301-307)(304-300)

(304-305)(304-307)×2.4829+(301-300)(301-304)(301-307)

(305-300)(305-304)(305-307)×2.4843+(301-300)(301-304)

(301-305)(307-300)(307-304)(307-305)×2.4871

$y(301)$=(-3)(-4)(-6)(-4)(-5)(-7)×2.4771+(1)(-4)(-6)(4)(-1)

(-3)×2.4829+(1)(-3)(-6)(5)(1)(-2)×2.4843+(1)(-3)

134

$(-4)(7)(3)(2)×2.4871$

$y(301)=-72-140×2.4771+2412×2.4829+18-10×2.4843+1242×2.4871$

$y(301)=2.4786$

Solution of the polynomial at point 301 is $y(301)=2.4786$

Example.3.17 Find Solution using Lagrange's Interpolation formula, find x=2.7.

x	f(x)
2	0.69315
2.5	0.91629
3	1.09861

Solution Given

x	2	2.5	3
y	0.69315	0.91629	1.09861

Lagrange's interpolating polynomial is

135

$$y = f(x) = \frac{(x-x_1)(x-x_2)....(x-x_n)}{(x_0-x_1)(x_0-x_2)....(x_0-x_n)} f(x_0)$$

$$+ \frac{(x-x_0)(x-x_2)....(x-x_n)}{(x_1-x_0)(x_1-x_2)....(x_1-x_n)} f(x_1)$$

$$+............\frac{(x-x_0)(x-x_1)....(x-x_{n-1})}{(x_n-x_0)(x_n-x_1)....(x_n-x_{n-1})} f(x_n)$$

$y(2.7) = (2.7\text{-}2.5)(2.7\text{-}3)(2\text{-}2.5)(2\text{-}3)\times0.69315+(2.7\text{-}2)$

$(2.7\text{-}3)(2.5\text{-}2)(2.5\text{-}3)\times0.91629+(2.7\text{-}2)(2.7\text{-}2.5)(3\text{-}2)$

$(3\text{-}2.5)\times1.09861$

$y(2.7) = (0.2)(\text{-}0.3)(\text{-}0.5)(\text{-}1)\times0.69315+(0.7)(\text{-}0.3)(0.5)$

$(\text{-}0.5)\times0.91629+(0.7)(0.2)(1)(0.5)\times1.09861$

$y(2.7)=\text{-}0.060.5\times0.69315+\text{-}0.21\text{-}0.25\times0.91629+0.140.5\times1.09861$

$y(2.7)=0.994116$

Solution of the polynomial at point 2.7 is $y(2.7)=0.994116$

Example.3.18 Find Solution using Lagrange's Interpolation formula, find x=1. Find f(2)

x	f(x)
-1	3
0	-6
3	39
6	822
7	1611

Solution Given that

x	-1	0	3	6	7
y	3	-6	39	822	1611

Lagrange's Interpolation formula is

$$y = f(x) = \frac{(x-x_1)(x-x_2)....(x-x_n)}{(x_0-x_1)(x_0-x_2)....(x_0-x_n)} f(x_0)$$

$$+ \frac{(x-x_0)(x-x_2)....(x-x_n)}{(x_1-x_0)(x_1-x_2)....(x_1-x_n)} f(x_1)$$

$$+............\frac{(x-x_0)(x-x_1)....(x-x_{n-1})}{(x_n-x_0)(x_n-x_1)....(x_n-x_{n-1})} f(x_n)$$

$y(1) = (1-0)(1-3)(1-6)(1-7)(-1-0)(-1-3)(-1-6)(-1-7)\times3+(1--1)$

$(1-3)(1-6)(1-7)(0--1)(0-3)(0-6)(0-7)\times-6+(1--1)(1-0)(1-6)(1-7)$

137

(3-1)(3-0)(3-6)(3-7)×39+(1--1)(1-0)(1-3)(1-7)(6--1)(6-0)(6-3)

(6-7)×822+(1--1)(1-0)(1-3)(1-6)(7--1)(7-0)(7-3)(7-6)×1611

$y(1) = (1)(-2)(-5)(-6)(-1)(-4)(-7)(-8)×3+(2)(-2)(-5)(-6)(1)$

$\quad (-3)(-6)(-7)×-6+(2)(1)(-5)(-6)(4)(3)(-3)(-4)×39+(2)(1)$

$\quad (-2)(-6)(7)(6)(3)(-1)×822+(2)(1)(-2)(-5)(8)(7)(4)(1)×1611$

$y(1) = (-0.2679)×3+0.9524×-6+0.4167×39$

$\quad +(-0.1905)×822+0.0893×1611$

$y(1)=-3$
Solution of the polynomial at point 1 is $y(1)=-3$

Example.3.19 Find Solution using Lagrange's Inverse Interpolation formula, x = 6. Find f(2).

x	f(x)
168	3
120	7
72	9
63	10

Solution: Given

x	168	120	72	63
y	3	7	9	10

Lagrange's interpolating polynomial

The value of x at you want to find $Pn(x):y=6$

Lagrange's Inverse Interpolation formula is

$$y = f(x) = \frac{(x-x_1)(x-x_2)....(x-x_n)}{(x_0-x_1)(x_0-x_2)....(x_0-x_n)} f(x_0)$$

$$+ \frac{(x-x_0)(x-x_2)....(x-x_n)}{(x_1-x_0)(x_1-x_2)....(x_1-x_n)} f(x_1)$$

$$+ .... \frac{(x-x_0)(x-x_1)....(x-x_{n-1})}{(x_n-x_0)(x_n-x_1)....(x_n-x_{n-1})} f(x_n)$$

y(6) = (6-7)(6-9)(6-10)(3-7)(3-9)(3-10)×168+(6-3)(6-9)(6-10)

(7-3)(7-9)(7-10)×120+(6-3)(6-7)(6-10)(9-3)(9-7)

(9-10)×72+(6-3)(6-7)(6-9)(10-3)(10-7)(10-9)×63

y(6) = (-1)(-3)(-4)(-4)(-6)(-7)×168+(3)(-3)(-4)(4)(-2)

(-3)×120+(3)(-1)(-4)(6)(2)(-1)×72+(3)(-1)

(-3)(7)(3)(1)×63

y(6) = 0.0714×168+1.5×120+(-1)×72+0.4286×63

y(6) = 147

Solution of the polynomial at point 6 is y(6)=147

3.6. Inverse lagrange's interpolation formula

Inverse lagrange's interpolation formula is obtained by interchanging x and $f(x)$ in lagrange's interpolation formula. The formula is

$$x = f(y) = \frac{(y-y_1)(y-y_2)....(y-y_n)}{(y_0-y_1)(y_0-y_2)....(y_0-y_n)} f(y_0)$$

$$+ \frac{(y-y_0)(y-y_2)....(y-y_n)}{(y_1-y_0)(y_1-y_2)....(y_1-y_n)} f(y_1)$$

$$+............\frac{(y-y_0)(y-y_1)....(y-y_{n-1})}{(y_n-y_0)(y_n-y_1)....(y_n-y_{n-1})} f(y_n)$$

Example.3.20   Find the value of x when y=0.3 by applying lagrange's formula inversely,

x	0.4	0.6	0.8
y	0.3683	0.3332	0.2897

Solution Here

$$x_0 = 0.4, x_1 = 0.6, x_2 = 0.8, y = 0.3$$
$$y_0 = 0.3683, y_1 = 0.3332, y_2 = 0.2897$$

By inverse lagrange's interpolation formula,

$$x = f(y) = \frac{(y-y_1)(y-y_2)....(y-y_n)}{(y_0-y_1)(y_0-y_2)....(y_0-y_n)} f(y_0)$$

$$+\frac{(y-y_0)(y-y_2)....(y-y_n)}{(y_1-y_0)(y_1-y_2)....(y_1-y_n)} f(y_1)$$

$$+..........\frac{(y-y_0)(y-y_1)....(y-y_{n-1})}{(y_n-y_0)(y_n-y_1)....(y_n-y_{n-1})} f(y_n)$$

$$x = \frac{(0.3-0.3332)(0.3-0.2897)}{(0.3683-0.3332)(0.3683-0.2897)} 0.4$$

$$+\frac{(0.3-0.3683)(0.3-0.2897)}{(0.3332-0.3683)(0.3332-0.2897)} 0.6$$

$$+\frac{(0.3-0.3683)(0.3-0.3332)}{(0.2897-0.3683)(0.2897-0.3332)} 0.8$$

$$\therefore x = 0.7574$$

3.7. Divided difference

In the construction of finite difference tables, x is assumed to be equally spaced. If x is not equally spaced, Lagrange's formula is used to find the unknown value from the table. If an another interpolation point is added to the tabulated data, then the Lagrangian coefficients are to be recalculated which results a different Lagrange's polynomial of higher degree. This difficulty will overcome by taking the Newton's divided differences.

Newton's divided difference formula is

$$f(x) = f(x_0) + (x-x_0)f(x_0,x_1) + (x-x_0)(x-x_1)$$
$$f(x_0,x_1,x_2) + ...$$

Newton's divided difference table is given below

141

X	Y	Δ	Δ^2	Δ^3	Δ^4
x_0	y_0				
x_1	y_1	$\Delta y_0 = \dfrac{y_1 - y_0}{x_1 - x_0}$			
x_2	y_2	$\Delta y_1 = \dfrac{y_2 - y_1}{x_2 - x_1}$	$\Delta^2 y_0 = \dfrac{\Delta y_1 - \Delta y_0}{x_2 - x_0}$		
x_3	y_3	$\Delta y_2 = \dfrac{y_3 - y_2}{x_3 - x_2}$	$\Delta^2 y_1 = \dfrac{\Delta y_2 - \Delta y_1}{x_3 - x_1}$	$\Delta^3 y_0 = \dfrac{\Delta^2 y_1 - \Delta^2 y_0}{x_3 - x_0}$	
x_4	y_4	$\Delta y_3 = \dfrac{y_4 - y_3}{x_4 - x_3}$	$\Delta^2 y_2 = \dfrac{\Delta y_3 - \Delta y_2}{x_4 - x_2}$	$\Delta^3 y_1 = \dfrac{\Delta^2 y_2 - \Delta^2 y_1}{x_4 - x_1}$	$\Delta^4 y_0 = \dfrac{\Delta^3 y_1 - \Delta^3 y_0}{x_4 - x_0}$

Examples.3.21 Find Solution using Newton's Divided Difference Interpolation formula

x	f(x)
300	2.4771
304	2.4829
305	2.4843
307	2.4871

Solution Given

x	300	304	305	307
y	2.4771	2.4829	2.4843	2.4871

Numerical divided differences method to find solution Newton's divided difference table is

x	y	1*st* order	2*nd* order
300	2.4771		
		2.4829-2.4771304-300=0.0014	
304	2.4829		0.0014-0.0014305-300=0
		2.4843-2.4829305-304=0.0014	
305	2.4843		0.0014-0.0014307-304=0
		2.4871-2.4843307-	

		305=0.0014	
307	2.4871		

The value of x at you want to find the $f(x)$:x=301

Newton's divided difference interpolation formula is

$$f(x) = f(x_0) + (x - x_0)f(x_0, x_1) + (x - x_0)(x - x_1)f(x_0, x_1, x_2) + \ldots$$

$y(301) = 2.4771 + (301-300) \times 0.0014 + (301-300)(301-304) \times 0$

$y(301) = 2.4771 + (1) \times 0.0014 + (1)(-3) \times 0$

$y(301) = 2.4771 + 0.0014 + 0$

$y(301) = 2.4785$

Solution of divided difference interpolation method $y(301) = 2.4785$

Examples.3.22 Find Solution using Newton's Divided Difference Interpolation formula, find x=2.7

x	f(x)
2	0.69315
2.5	0.91629
3	1.09861

Solution The value of table for x and y

x	2	2.5	3
y	0.6932	0.9163	1.0986

Numerical divided differences method to find solution

Newton's divided difference table is

x	y	1st order	2nd order
2	0.6932		
		0.9163-0.69322.5-2 =0.4463	
2.5	0.9163		0.3646-0.44633-2 = -0.0816
		1.0986-0.91633-2.5 =0.3646	
3	1.0986		

The value of x at you want to find the $f(x):x=2.7$

Newton's divided difference interpolation formula is

145

$$f(x) = f(x_0) + (x - x_0) f(x_0, x_1)$$
$$+ (x - x_0)(x - x_1) f(x_0, x_1, x_2) + \ldots$$

$y(2.7)=0.6932+(2.7-2)\times0.4463+(2.7-2)(2.7-2.5)\times0.0816$

$y(2.7)=0.6932+(0.7)\times0.4463+(0.7)(0.2)\times-0.0816$

$y(2.7)=0.6932+0.3124-0.0114$

$y(2.7)=0.9941$

Solution of divided difference interpolation method $y(2.7)=0.9941$.

Examples.3.23 Find Solution of an equation $x^3 - x + 1$ using Newton's Divided Difference Interpolation formula, $x_1 = 2$ and $x_2 = 4, x = 3.8$, Step value (h) = 0.5, finding f(2). **Solution** Equation is $f(x) = x^3 - x + 1$.

x	2	2.5	3	3.5	4
y	7	14.125	25	40.375	61

Numerical divided differences method to find solution

Newton's divided difference table is

x	y	1st order	2nd order	3rd order	4th order
2	7				
		14.25			

146

2.5	14.125		7.5		
		21.75		1	
3	25		9		0
		30.75		1	
3.5	40.375		10.5		
		41.25			
4	61				

The value of x at you want to find the $f(x)$: $x=3.8$

Newton's divided difference interpolation formula is

$$f(x) = f(x_0) + (x - x_0) f(x_0, x_1)$$
$$+ (x - x_0)(x - x_1) f(x_0, x_1, x_2) + \ldots$$

$y(3.8) = 7 + (3.8-2) \times 14.25 + (3.8-2)(3.8-2.5) \times 7.5 + (3.8-2)$
$(3.8-2.5)(3.8-3) \times 1 + (3.8-2)(3.8-2.5)(3.8-3)(3.8-3.5) \times 0$
$y(3.8) = 7 + (1.8) \times 14.25 + (1.8)(1.3) \times 7.5 + (1.8)(1.3)(0.8) \times 1$
$+ (1.8)(1.3)(0.8)(0.3) \times 0$
$y(3.8) = 7 + 25.65 + 17.55 + 1.872 + 0$
$y(3.8) = 52.072$
Solution of divided difference interpolation
method $y(3.8) = 52.072$

Examples.3.24 Find Solution of an equation $2x^3 - 4x + 1$ using Newton's Divided Difference Interpolation formula, $x_1 = 2$ and $x_2 = 4, x = 3.8$, Step value (h) = 0.5, find f(2)

Solution Equation is $f(x) = 2x^3 - 4x + 1$.

The value of table for x and y

x	2	2.5	3	3.5	4
y	9	22.25	43	72.75	113

Numerical divided differences method to find solution

Newton's divided difference table is

x	y	1st order	2nd order	3rd order	4th order
2	9				
		26.5			
2.5	22.25		15		
		41.5		2	
3	43		18		0
		59.5		2	
3.5	72.75		21		

148

		80.5			
4	113				

The   value   of x at   you   want   to   find   the $f(x)$:x=3.8

Newton's   divided   difference   interpolation   formula   is

$$f(x) = f(x_0) + (x - x_0) f(x_0, x_1)$$
$$+ (x - x_0)(x - x_1) f(x_0, x_1, x_2) + \dots$$

$y(3.8) = 9+(3.8-2)\times26.5+(3.8-2)(3.8-2.5)\times15+(3.8-2)$
(3.8-2.5)(3.8-3)×2+(3.8-2)(3.8-2.5)(3.8-3)(3.8-3.5)×0
$y(3.8)$=9+(1.8)×26.5+(1.8)(1.3)×15+(1.8)(1.3)(0.8)×2+(1.8)
(1.3)(0.8)(0.3)×0
$y(3.8)$=9+47.7+35.1+3.744+0
$y(3.8)$=95.544.

Solution   of   divided   difference   interpolation
method $y(3.8)$=95.544.

3.8. Stirling's formula

Gauss forward difference formula is given by

$$y = y_0 + p\Delta y_0 + \frac{p(p-1)}{\angle 2} \Delta^2 y_{-1} + \frac{p(p-1)(p+1)}{\angle 3} \Delta^3 y_{-1} + \dots$$

Gauss backward difference formula is given by

149

$$y = y_0 + p\Delta y_{-1} + \frac{p(p-1)}{\angle 2}\Delta^2 y_{-1} + \frac{p(p-1)(p+1)}{\angle 3}\Delta^3 y_{-2}$$

$$+\ldots\ldots$$

We will get Stirling's formula.

$$\frac{y+y}{2} = \frac{y_0+y_0}{2} + \frac{p\Delta y_0 + p\Delta y_{-1}}{2} + \frac{1}{2}\left[\frac{p(p-1)}{\angle 2}\Delta^2 y_{-1} + \frac{p(p+1)}{\angle 2}\Delta^2 y_{-1}\right]$$

$$+\frac{1}{2}\left[\frac{p(p-1)(p-2)(p+1)}{\angle 4}\Delta^4 y_{-2} + \frac{p(p-1)(p+2)(p+1)}{\angle 4}\Delta^4 y_{-2}\right]+\ldots.$$

$$\Rightarrow y = y_0 + \frac{p}{2}\left(\Delta y_0 + \Delta y_{-1}\right) + \frac{p}{2}\left[\frac{p-1}{2}+\frac{p+1}{2}\right]\Delta^2 y_{-1}$$

$$+\frac{p(p-1)(p+1)}{6}\left(\Delta^3 y_{-1} + \Delta^3 y_{-2}\right)$$

$$+\frac{1}{2}\frac{p(p-1)(p+1)}{24}\left[p-2+p+2\right]\Delta^4 y_{-2}+\ldots\ldots$$

This is stirring's formula, it gives the most accurate result for $-0.25 \le p \le 0.25$.

Therefore, we have to choose x_0 such that p satisfies this inequality.

Examples.3.25. Using Stirring's formula to find y_{28},

given $\quad y_{20} = 49225, y_{25} = 48316, y_{30} = 47236,$
$y_{35} = 45926, y_{40} = 44306$

Solution

x	Y	Δ	Δ^2	Δ^3	Δ^4
20	49225				
25	48316	-909			
30	47236	-1080	-171		
35	45926	-1310	-230	-59	
40	44306	-1620	-310	-80	-21

Here $x = 28, x_0 = 30, h = 5, p = \dfrac{x - x_0}{h} = \dfrac{-2}{5} = -0.4$

By, Stirling's formula,

$$\Rightarrow y = y_0 + \frac{p}{2}\left(\Delta y_0 + \Delta y_{-1}\right) + \frac{p}{2}\left[\frac{p-1}{2} + \frac{p+1}{2}\right]\Delta^2 y_{-1}$$

$$+\frac{p(p-1)(p+1)}{6}\left(\Delta^3 y_{-1} + \Delta^3 y_{-2}\right)$$

$$+\frac{1}{2}\frac{p(p-1)(p+1)}{24}\left[p-2+p+2\right]\Delta^4 y_{-2} + \ldots\ldots$$

$$\Rightarrow y = 47236 + \frac{-0.4}{2}\left[-1310 - 1080\right]$$

$$+\frac{(-0.4)^2}{2}(-230) + \frac{-0.4}{3}\frac{\left((-0.4)^2 - 1\right)}{2}\left(\frac{-80-59}{2}\right)$$

$$+\frac{-0.4}{24}\frac{\left((-0.4)^2 - 1\right)}{24}(-21)$$

$$y = 47691.82$$

151

Examples.3.26. Using Stirling's formula to find y_{32} from the

following table.
$$y_{20} = 14.035, y_{25} = 13.674, y_{30} = 13.257,$$
$$y_{35} = 12.734, y_{40} = 12.089, y_{45} = 11.309$$

Solution The difference table is given below

X	Y	Δ	Δ^2	Δ^3	Δ^4	Δ^5
20	14.035					
25	13.674	-0.361				
30	13.257	-0.417	-0.056			
35	12.734	-0.523	-0.106	-0.05		
40	12.089	-0.645	-0.122	-0.016	0.034	
45	11.309	-0.78	-0.135	-0.013	0.003	-0.031

Here $x = 32, x_0 = 30, h = 5, p = \dfrac{x - x_0}{h} = \dfrac{2}{5} = 0.4$

By, Stirling's formula,

$$y = y_0 + \frac{p}{2}\left(\Delta y_0 + \Delta y_{-1}\right) + \frac{p}{2}\left[\frac{p-1}{2} + \frac{p+1}{2}\right]\Delta^2 y_{-1}$$

$$+ \frac{p(p-1)(p+1)}{6}\left(\Delta^3 y_{-1} + \Delta^3 y_{-2}\right)$$

$$+ \frac{1}{2}\frac{p(p-1)(p+1)}{24}[p-2+p+2]\Delta^4 y_{-2} + \ldots\ldots$$

$$\Rightarrow y = 13.257 + \frac{0.4}{2}[-0.417 - 0.523] + \frac{(0.4)^2}{2}(-0.106)$$

$$+ \frac{0.4}{3}\frac{\left((0.4)^2 - 1\right)}{2}\left(\frac{-0.05 - 0.016}{2}\right) + \frac{-0.4}{24}\frac{\left((0.4)^2 - 1\right)}{24}(0.034)$$

$$= 13.257 - 0.188 - 0.00848 + 0.001848 - 0.0001904$$

$$y = 13.062$$

Examples.3.27

Find Solution using Stirling's formula, find x=28.

x	f(x)
20	49225
25	48316
30	47236
35	45926
40	44306

Solution Given

x	20	25	30	35	40
y	49225	48316	47236	45926	44306

Stirling's method to find solution $h = 25-20 = 5$

Taking $x_1 = 30$ then $p = \dfrac{x - x_0}{h} = \dfrac{x - 30}{5}$

The difference table is

x	p	y	Δy	Δ2y	Δ3y	Δ4y
20	-2	49225				
			-909			
25	-1	48316		-171		
			-1080		-59	
30	0	47236		-230		-21
			-1310		-80	
35	1	45926		-310		
			-1620			
40	2	44306				

x=28,p=-0.4

$y_0 = 47236, \Delta y_0 = -1310, \Delta^2 y_{-1} = -230,$

$\Delta^3 y_{-1} = -80, \Delta^4 y_{-2} = -21$

154

Stirling's formula,

$$\frac{y+y}{2} = \frac{y_0+y_0}{2} + \frac{p\Delta y_0 + p\Delta y_{-1}}{2} + \frac{1}{2}\left[\frac{p(p-1)}{\angle 2}\Delta^2 y_{-1} + \frac{p(p+1)}{\angle 2}\Delta^2 y_{-1}\right]$$
$$+\frac{1}{2}\left[\frac{p(p-1)(p-2)(p+1)}{\angle 4}\Delta^4 y_{-2} + \frac{p(p-1)(p+2)(p+1)}{\angle 4}\Delta^4 y_{-2}\right]+....$$

y-0.4=47236+(-0.4)·(-1310-1080)2+(0.16)2·(-230)

   +(-0.4)(0.16-1)6·(-80-59)2+(0.16)(0.16-1)24·(-21)

y-0.4=47236+478-18.4-3.892+0.1176

y-0.4=47691.8256
The solution of Stirring's interpolation is $y(28)$=47691.8256

Examples.3.28. Find Solution using Stirling's formula

x	f(x)
10	0.23967
11	0.28060
12	0.31788
13	0.35209
14	0.38368

find x=12.2.

Solution Given

x	10	11	12	13	14
y	0.23967	0.2806	0.31788	0.35209	0.38368

Stirling's method to find solution

h=11-10=1. Taking x_0=12 then p=x-121

The difference table is

x	p=x-121	y	Δy	$\Delta 2y$	$\Delta 3y$	$\Delta 4y$
10	-2	0.23967				
			0.04093			
11	-1	0.2806		-0.00365		
			0.03728		0.00058	
12	0	0.31788		-0.00307		-0.00013
			0.03421		0.00045	
13	1	0.35209		-0.00262		
			0.03159			
14	2	0.38368				

$x=12.2$

$$p = \frac{x - x_0}{h} = \frac{12.2 - 12}{1} = 0.2$$

$$y_0 = 0.31788, \Delta y_0 = 0.03421, \Delta^2 y_{-1} = -0.00307,$$

$$\Delta^3 y_{-1} = 0.00045, \Delta^4 y_{-2} = -0.00013$$

Stirling's formula,

$$\frac{y + y}{2} = \frac{y_0 + y_0}{2} + \frac{p\Delta y_0 + p\Delta y_{-1}}{2}$$

$$+ \frac{1}{2}\left[\frac{p(p-1)}{\angle 2} \Delta^2 y_{-1} + \frac{p(p+1)}{\angle 2} \Delta^2 y_{-1} \right]$$

$$+ \frac{1}{2}\left[\begin{array}{c} \dfrac{p(p-1)(p-2)(p+1)}{\angle 4} \Delta^4 y_{-2} \\ + \dfrac{p(p-1)(p+2)(p+1)}{\angle 4} \Delta^4 y_{-2} \end{array} \right] + ...$$

$y_{0.2}$=0.31788+(0.2)·(0.03421+0.03728)2+(0.04)2

(-0.00307)+(0.2)(0.04-1)6·(0.00045+0.00058)2

+(0.04)(0.04-1)24·(-0.00013)

$y_{0.2}$=0.31788+0.007149-0.0000614-0.00001648+0.0000208

$y_{0.2}$=0.324951

The solution of Stirring's interpolation is $y(12.2)$=0.324951

Example.3.29 Find the Solution using stirring's formula, x=16, finding f(2).

x	f(x)
0	0
5	0.0875
10	0.1763
15	0.2679
20	0.3640
25	0.4663
30	0.5774

Solution Given that

x	0	5	10	15	20	25	30
y	0	0.0875	0.1763	0.2679	0.364	0.4663	0.5774

Stirling's method to find solution $h=5-0=5$

Taking $x_0=15$ then $p=x-155$.   The difference table is

x	$p=$ $x\text{-}155$	y	Δy	$\Delta 2y$	$\Delta 3y$	$\Delta 4y$	$\Delta 5y$	$\Delta 6y$
0	-3	0						
			0.0875					
5	-2	0.0875		0.0013				
			0.0888		0.0015			
10	-1	0.1763		0.0028		0.0002		
			0.0916		0.0017		-0.0002	
15	0	0.2679		0.0045		0		0.0011
			0.0961		0.0017		0.0009	
20	1	0.364		0.0062		0.0009		
			0.102		0.002			

159

			3		6			
25	2	0.466 3		0.008 8				
			0.111 1					
30	3	0.577 4						

$$x=16. p=0.2$$

$$y_0 = 0.2679, \Delta y_0 = 0.0961, \Delta^2 y_{-1} = 0.0045, \Delta^3 y_{-1} = 0.0017,$$

$$\Delta^4 y_{-2} = 0, \Delta^5 y_{-2} = 0.0009, \Delta^6 y_{-3} = 0.0011$$

Stirling's formula,

$$\frac{y+y}{2} = \frac{y_0+y_0}{2} + \frac{p\Delta y_0 + p\Delta y_{-1}}{2}$$

$$+\frac{1}{2}\left[\frac{p(p-1)}{\angle 2}\Delta^2 y_{-1} + \frac{p(p+1)}{\angle 2}\Delta^2 y_{-1}\right]$$

$$+\frac{1}{2}\left[\begin{array}{c}\dfrac{p(p-1)(p-2)(p+1)}{\angle 4}\Delta^4 y_{-2} \\ +\dfrac{p(p-1)(p+2)(p+1)}{\angle 4}\Delta^4 y_{-2}\end{array}\right]$$

$$+....$$

$y_{0.2}$=0.2679+(0.2)·(0.0961+0.0916)2+(0.04)2·(0.0045)+(0.2)

(0.041)6·(0.0017+0.0017)2+(0.04)(0.04-1)24·(0)+(0.2)

(0.04-1)(0.04-4)120·(0.0009)2+(0.04)(0.04-1)(0.04-4)720·(0.0011)

160

$y_{0.2} = 0.2679+0.01877+0.00009-$
$0.0000544+0+0.0000022176+0.0000002323$
$y_{0.2}=0.2867$

Solution of stirring's interpolation is $y(16)=0.2867$

3.9. Linear Spline interpolation

In the mathematical field of numerical analysis, **spline interpolation** is a form of interpolation where the interpolant is a special type of piecewise polynomial called a spline. Spline interpolation is preferred over polynomial interpolation because the interpolation error can be made small even when using low degree polynomials for the spline. Thus the spline interpolation avoids the problem of Runge's phenomenon which occurs when using high degree polynomials. This types of splines are called basic splines i.e., B-splines.

Definition Given n+1 distinct points x_i , such that

with n+1 points y, we are trying to find a spline function of degree n

$$S(x) = \begin{cases} S_0(x), x \in [x_0, x_1] \\ S_1(x), x \in [x_1, x_2] \\ \dotsb \\ S_{n-1}(x), x \in [x_{n-1}, x_n] \end{cases}$$

where each $S(x)$ is a polynomial of degree k.

Using polynomial interpolation, the polynomial of degree n which interpolates the data set is uniquely defined by the data points. The spline of degree n which interpolates the same data set is not uniquely defined, and we have to fill in n-1 additional degrees of freedom to construct a unique spline interpolant.

161

Linear spline interpolation is the simplest form of spline interpolation and is equivalent to linear interpolation. The data points are graphically connected by straight lines. The resultant spline

would be a polygon if the end point is connected to the beginning points.

Algebraically, each S_i is a linear function constructed as

$$S_i(x) = y_i + \frac{y_{i+1} - y_i}{x_{i+1} - x_i}(x - x_i)$$

The spline must be continuous at each data point, that is

$$S_{i-1}(x_i) = S_i(x_i), i = 1, 2, \ldots n-1$$

This is the case as we can easily see

$$S_{i-1}(x_i) = y_{i-1} + \frac{y_i - y_{i-1}}{x_i - x_{i-1}}(x_i - x_{i-1}) = y_i$$

$$S_i(x_i) = y_i + \frac{y_{i+1} - y_i}{x_{i+1} - x_i}(x_i - x_i) = y_i$$

3.10. Cubic spline interpolation

For a data set $\{x_i\}$ of n+1 points, we can construct a cubic spline with n piecewise cubic polynomials between the data points, if

$$S(x) = \begin{cases} S_0(x), x \in [x_0, x_1] \\ S_1(x), x \in [x_1, x_2] \\ \ldots\ldots\ldots\ldots\ldots\ldots \\ S_{n-1}(x), x \in [x_{n-1}, x_n] \end{cases}$$

162

represents the spline function interpolating the function f, we require

i. The interpolating property, $S(x_i) = f(x_i)$

ii. The splines to join up, $S_{i-1}(x_i) = S_i(x_i), i = 1, 2, \ldots n-1$

iii. Twice continuous differentiable, $S_{i-1}(x_i) = S_i(x_i)$, and $S''_{i-1}(x_i) = S''_i(x_i), i = 1, 2, \ldots n$

For the n cubic polynomials comprising S, we need to determine 4n conditions to determine these polynomials. However, the interpolating property gives us n+1 conditions, and the conditions on the interior data points give us n+1-2=n-1 data points each, summing to 4n-2 conditions. We require two other conditions, and these can be imposed upon the problem for different reasons. One such choice results in the so-called clamped cubic spline, with $S'(x_0) = u, S'(x_n) = v$ for given values u and v.

Alternatively, we can set $S''(x_0) = S''(x_n) = 0$

Resulting in the natural cubic spline. The natural cubic spline is approximately the same curve as created by the spline device.

Amongst all twice continuously differentiable functions, clamped and natural cubic splines yield the least oscillation about the function f which is interpolated.

Another choice gives the periodic cubic spline if

$$S(x_0) = S(x_n)$$
$$S'(x_0) = S'(x_n)$$
$$S''(x_0) = S''(x_n)$$

Another choice gives the complete cubic spline if

$$S(x_0) = S(x_n)$$
$$S'(x_0) = S'(x_n)$$
$$S''(x_0) = f'(x_0), S''(x_n) = f'(x_n)$$

Now we discuss the following method regarding to the cubic-spline interpolation.

3.11. The cubic-spline method

Consider the function $y = f(x)$ in [a,b], divide the interval [a,b] into n subintervals of length h, $a = x_0, x_1, x_2, \ldots x_{i-1}, x_i \ldots x_n = b$. If $S(x)$ is the cubic spline in (x_{i-1}, x_i), then we have $\int_{x_0}^{x_n} y dx = \sum_{i=1}^{n} \int_{x_{i-1}}^{x_i} s_i(x) dx \ldots \ldots (i)$

The cubic spline for each interval (x_{i-1}, x_i) is given by

$$S_i(x) = \frac{(x_i - x)^3}{6(x_i - x_{i-1})} s''(x_{i-1}) + \frac{(x - x_{i-1})^3}{6(x_i - x_{i-1})} s''(x_i)$$

$$+ \frac{x_i - x}{x_i - x_{i-1}} \left[s(x_{i-1}) - \frac{(x_i - x_{i-1})^2}{6} s''(x_{i-1}) \right]$$

$$+ \frac{x - x_{i-1}}{x_i - x_{i-1}} \left[s(x_i) - \frac{(x_i - x_{i-1})^2}{6} s''(x_i) \right]$$

$$S(x) = \frac{(x_i - x)^3}{6h} s''(x_{i-1}) + \frac{(x - x_{i-1})^3}{6h} s''(x_i) + \frac{x_i - x}{h}$$
$$\left(s(x_{i-1}) - \frac{h^2}{6} s''(x_{i-1}) \right)$$
$$+ \frac{x - x_{i-1}}{h} \left(s(x_i) - \frac{h^2}{6} s''(x_i) \right)$$

164

Substituting these values

$$\int_{x_0}^{x_n} y dx = \sum_{i=1}^{n} \int_{x_{i-1}}^{x_i} \frac{1}{6h} \left[(x_i - x)^3 M_{i-1} + (x - x_{i-1})^3 M_i \right]$$

$$+ \frac{x_i - x}{h} \left(y_{i-1} - \frac{h^2}{6} M_{i-1} \right)$$

$$+ \frac{1}{h} (x - x_{i-1}) \left[y_i - \frac{h^2}{6} M_i \right] dx$$

$$= \sum_{i=1}^{n} \frac{h}{2} (y_{i-1} + y_i) - \frac{h^3}{24} (M_{i-1} + M_i) \quad \text{on simplification where}$$

$M_{i-1} = s''(x_{i-1}), M_i = s''(x_i)$ and these are calculated from the recurrence relation.

$$M_{i-1} + 4M_i + M_{i+1} = \frac{6}{h^2} (y_{i-1} - 2y_i + y_{i+1}), \, for \, i = 1, 2, ... n - 1$$

Example.3.30 Evaluate $\int_{0}^{1} \cos x dx$ using cubic spline method

Solution Let

X	$x_0 = 0$	$x_1 = 0.5$	$x_2 = 1$
$y = \cos x$	$y_0 = 1$	$y_1 = 0.8776$	$y_2 = 0.5403$

Put n=2, $h = \dfrac{b-a}{n} = \dfrac{1-0}{2} = 0.5$

The recurrence relation is

$$M_{i-1} + 4M_i + M_{i+1} = \frac{6}{h^2} (y_{i-1} - 2y_i + y_{i+1})$$

165

Put $i = 1 \Rightarrow M_0 + 4M_1 + M_2 = \dfrac{6}{h^2}(y_0 - 2y_1 + y_2)$

We have $M_0 = y_0''$, $M_n = y_n'' \Rightarrow M_2 = y_2''$

$y' = -\sin x \Rightarrow y'' = -\cos x \Rightarrow y_0'' - \cos 0 = -1$

$y'' = -\cos x \Rightarrow y_2'' = -0.5403$

Put in these, we get

$-1 + 4M_1 - 0.5403 = \dfrac{6}{0.5^2}\left(1 - 2(0.8776) + 0.5403\right)$

$\Rightarrow 4M_1 - 1.5403 = -5.1576$

$\Rightarrow M_1 = -0.9043$

Using cubic spline method,

$\displaystyle\int_0^1 \cos x\,dx = \sum_{i=1}^{2} \dfrac{h}{2}(y_{i-1} + y_i) - \dfrac{h^3}{24}(M_{i-1} + M_i)$

$= \dfrac{h}{2}(y_0 + y_1) - \dfrac{h^3}{24}(M_0 + M_1) + \dfrac{h}{2}(y_1 + y_2) - \dfrac{h^3}{24}(M_1 + M_2)$

$= \dfrac{h}{2}(y_0 + 2y_1 + y_2) - \dfrac{h^3}{24}(M_0 + 2M_1 + M_2)$

$= \dfrac{0.5}{2}(1 + 1.7552 + 0.5403) - \dfrac{0.125}{24}(-1 - 1.086 - 0.5403)$

$= 0.8413$

By direct integration, $\displaystyle\int_0^1 \cos x\,dx = 0.8415$

Absolute error = exact error-obtained error

$= 0.8415 - 0.8413 = 0.0002.$

Example.3.31 Calculate Cubic Splines

X	1	2	3	4
Y	1	5	11	8

y(1.5) and y'(2)

Solution Let

x	1	2	3	4
y	1	5	11	8

Cubic spline formula is

$$\int_{x_0}^{x_n} y\,dx = \sum_{i=1}^{n} \int_{x_{i-1}}^{x_i} \frac{1}{6h}\left[(x_i-x)^3\,M_{i-1}+(x-x_{i-1})^3\,M_i\right]$$
$$+\frac{x_i-x}{h}\left(y_{i-1}-\frac{h^2}{6}M_{i-1}\right)+\frac{1}{h}(x-x_{i-1})\left[y_i-\frac{h^2}{6}M_i\right]dx$$

We have,

$$M_{i-1}+4M_i+M_{i+1}=\frac{6}{h^2}\left(y_{i-1}-2y_i+y_{i+1}\right),\,for\,i=1,2,...n-1$$

Here $h=1, n=3$

$$M_0 = 0, M_3 = 0$$

Substitute i=1 in equation (2)

$$M_0 + 4M_1 + M_2 = 6h_2(y_0 - 2y_1 + y_2)$$
$$\Rightarrow 0 + 4M_1 + M_2 = 61 \cdot (1 - 2 \cdot 5 + 11)$$
$$\Rightarrow 4M_1 + M_2 = 12$$

Substitute i=2 in equation (2)

$$M_1 + 4M_2 + M_3$$
$$= 6h_2(y_1 - 2y_2 + y_3)$$
$$\Rightarrow M_1 + 4M_2 + 0$$
$$= 61 \cdot (5 - 2 \cdot 11 + 8) \Rightarrow M_1 + 4M_2 = -54$$

Solving these 2 equations using elimination method
Substitute i=1 in equation (1), we get cubic spline
in 1*st* interval [$x0,x1$]=[1,2]

$$f_1(x) = (x_1 - x)36hM_0 + (x - x_0)36hM_1$$
$$+(x_1 - x)h(y_0 - \frac{h_2}{6}M_0)$$
$$+(x - x_0)h\left(y_1 - \frac{h_2}{6}M_1\right)f_1(x)$$
$$= (2 - x)36 \cdot 0 + (x - 1)36 \cdot 6.8$$
$$+(2 - x)1(1 - 16 \cdot 0)$$
$$+(x - 1)1(5 - 16 \cdot 6.8)f_1(x)$$
$$= 1.1333x_3 - 3.4x_2$$
$$+6.2667x - 3, for\ 1 \leq x \leq 2$$

Substitute i=2 in equation (1), we get cubic spline
in 2*nd* interval [$x1,x2$]=[2,3]

$$f_2(x) = (x_2 - x)36hM_1 + (x - x_1)36hM_2$$

$$+(x_2 - x)h(y_1 - \frac{h_2}{6}M_1) + (x - x_1)h(y_2 - \frac{h_2}{6}M_2)f_2(x)$$

$$= (3 - x)36 \cdot 6.8 + (x - 2)36 \cdot -15.2 + (3 - x)1(5 - 16 \cdot 6.8)$$

$$+(x - 2)1(11 - 16 \cdot -15.2)f2(x)$$

$$= -3.6667x_3 + 25.4x_2$$

$$-51.3333x + 35.4, \text{ for } 2 \leq x \leq 3$$

Substitute i=3 in equation (1), we get cubic spline in 3rd interval $[x_2,x_3]$=[3,4]

$$f_3(x) = (x_3 - x)36hM_2 + (x - x_2)36hM_3 + (x_3 - x)$$

$$h(y_2 - \frac{h_2}{6}M_2) + (x - x_2)h(y_3 - \frac{h_2}{6}M_3)f_3(x)$$

$$= (4 - x)36 - 15.2 + (x - 3)36 \cdot 0 + (4 - x)1(11 - 16 \cdot -15.2)$$

$$+(x - 3)1(8 - 16 \cdot 0)f3(x)$$

$$= 2.5333x_3 - 30.4x_2 + 116.0667x - 132, \text{ for } 3 \leq x \leq 4$$

For $y(1.5)$, $1.5 \in [1,2]$, so substitute x=1.5 in $f1(x)$, we get

$$f_1(1.5) = 2.575$$

For $y'(2)$, $2 \in [1,2]$, So, find $f_1(1.5)$ i.e.,

$$f'_1(x) = 3.4x_2 - 6.8x + 6.2667$$

Now substitute x=2 in $f'1(x)$, we get

$$f'_1(2) = 6.2667.$$

Example.3.32. Calculate Cubic Splines

X	1	2	3	4
Y	1	2	5	11

y(1.5) and y'(3)

Solution Let

x	1	2	3	4
y	1	2	5	11

The cubic spline formula is

$$\int_{x_0}^{x_n} y\,dx = \sum_{i=1}^{n} \int_{x_{i-1}}^{x_i} \frac{1}{6h}\left[(x_i - x)^3 M_{i-1} + (x - x_{i-1})^3 M_i\right]$$

$$+ \frac{x_i - x}{h}\left(y_{i-1} - \frac{h^2}{6}M_{i-1}\right) + \frac{1}{h}(x - x_{i-1})\left[y_i - \frac{h^2}{6}M_i\right]dx$$

$$M_{i-1} + 4M_i + M_{i+1} = \frac{6}{h^2}\left(y_{i-1} - 2y_i + y_{i+1}\right),\, for\, i = 1, 2,\dots n-1$$

Here $h=1, n=3$

$$M_0 = 0, M_3 = 0$$

Substitute $i=1$ in

$M_0 + 4M_1 + M_2$

$= \dfrac{6}{h^2}(y_0 - 2y_1 + y_2)$

$\Rightarrow 0 + 4M_1 + M_2 = 61 \cdot (1 - 2 \cdot 2 + 5)$

$\Rightarrow 4M_1 + M_2 = 12$

Substitute $i=2$ in

$$M_1 + 4M_2 + M_3$$

$$= \dfrac{6}{h^2}(y_1 - 2y_2 + y_3)$$

$$\Rightarrow M_1 + 4M_2 + 0$$

$$= 61(2 - 2 \cdot 5 + 11)$$

$$\Rightarrow M_1 + 4M_2 = 18$$

Solving these two equations using the elimination method

Substitute $i=1$ in equation, we get cubic spline
in $1st$ interval $[x_0, x_1] = [1,2]$

$$f_1(x) = (x_1 - x)36hM_0 + (x - x_0)36hM_1 + (x_1 - x)h(y_0 - \dfrac{h^2}{6}M_0)$$

$$+ (x - x_0)h(y_1 - \dfrac{h^2}{6}M_1)f_1(x)$$

$$= (2 - x)36 \cdot 0 + (x - 1)36 \cdot 2 + (2 - x)1(1 - 16 \cdot 0)$$

$$+ (x - 1)1(2 - 16 \cdot 2)f_1(x)$$

$$= 13(x_3 - 3x_2 + 5x), \; for \, 1 \le x \le 2$$

Substitute $i=2$ in equation, we get cubic spline

in 2nd interval $[x_1,x_2]=[2,3]$

$$f_2(x)=(x_2-x)\frac{36}{h}M_1+(x-x_1)\frac{36}{h}M_2$$

$$+(x_2-x)h(y_1-\frac{h^2}{6}M_1)$$

$$+(x-x_1)h(y_2-\frac{h^2}{6}M_2)f_2(x)$$

$$=(3-x)36\cdot2+(x-2)36\cdot4+(3-x)1(2-16\cdot2)$$

$$+(x-2)1(5-16\cdot4)f_2(x)$$

$$=13(x_3-3x_2+5x),\ for\ 2\le x\le3$$

Substitute $i=3$ in   equation (1),   we   get   cubic   spline in 3rd interval $[x2,x3]=[3,4]$

$$f_3(x)=(x_3-x)\frac{36}{h}M_2+(x-x_2)\frac{36}{h}M_3+(x_3-x)h(y_2-\frac{h^2}{6}M_2)$$

$$+(x-x_2)h(y_3-\frac{h^2}{6}M_3)f_3(x)$$

$$=(4-x)36\cdot4+(x-3)36\cdot0+(4-x)1(5-16\cdot4)+(x-3)$$

$$(11-16\cdot0)f_3(x)$$

$$=13(-2x_3+24x_2-76x+81),\ for\ 3\le x\le4$$

For $y(1.5)$, $1.5\in[1,2]$,

 substitute $x=1.5$ in $f_1(x)$, we get

$$f'(1.5)=1.375$$
For $y'(3)$, $3\in[2,3]$,so, find $f'_2(x)$

$$f'_2(x)=13(3x_2-6x+5)$$
 Now, substitute $x=3$ in $f'_2(x)$,we get

172

$$f'2(3)=4.6667$$

Example.3.33 Calculate Cubic Splines

X	0	1	2
Y	-5	-4	3

Find y(0.5)

Solution   Cubic spline formula is

$$\int_{x_0}^{x_n} y\,dx = \sum_{i=1}^{n} \int_{x_{i-1}}^{x_i} \frac{1}{6h}\left[(x_i-x)^3 M_{i-1} + (x-x_{i-1})^3 M_i\right] + \frac{x_i-x}{h}\left(y_{i-1} - \frac{h^2}{6} M_{i-1}\right)$$
$$+\frac{1}{h}(x-x_{i-1})\left[y_i - \frac{h^2}{6} M_i\right]dx$$

$$M_{i-1} + 4M_i + M_{i+1} = \frac{6}{h^2}\left(y_{i-1} - 2y_i + y_{i+1}\right), for\, i = 1, 2, ...n-1$$

Here $h=1, n=2$ $M_0=0, M_2=0$

Substitute $i=1$ in the equation.

$$M_0 + 4M_1 + M_2$$

$$= \frac{6}{h^2}(y_0 - 2y_1 + y_2)$$

$$\Rightarrow 0 + 4M_1 + 0$$

$$= 61 \cdot (-5 - 2 \cdot -4 + 3)$$

$$\Rightarrow 4M_1 = 36 \Rightarrow M_1 = 9$$

Substitute $i=1$ in equation (1), we get cubic spline in $1st$ interval $[x0,x1]=[0,1]$

$$f_1(x) = (x_1 - x)36hM_0 + (x - x_0)36hM_1$$

$$+(x_1 - x)h(y_0 - \frac{h^2}{6}M_0) + (x - x_0)h(y_1 - \frac{h^2}{6}M_1)f_1(x)$$

$$= (1-x)36 \cdot 0 + (x-0)36 \cdot 9 + (1-x)1(-5 - 16 \cdot 0)$$

$$+(x-0)1(-4 - 16 \cdot 9)f1(x)$$

$$= 12(3x_3 - x - 10), \, for \, 0 \le x \le 1$$

Substitute $i=2$ in equation (1), we get cubic spline in $2nd$ interval $[x_1,x_2]=[1,2]$.

$$f_2(x) = (x_2 - x)\frac{36}{h}M_1 + (x - x_1)\frac{36}{h}M_2$$

$$+(x_2 - x)h(y_1 - \frac{h^2}{6}M_1) + (x - x_1)h(y_2 - \frac{h^2}{6}M_2)f_2(x)$$

$$= (2-x)36 \cdot 9 + (x-1)36 \cdot 0 + (2-x)1(-4 - 16 \cdot 9) + (x-1)$$

$$1(3 - 16 \cdot 0)f_2(x)$$

$$= 12(-3x_3 + 18x_2 - 19x - 4), \, for \, 1 \le x \le 2$$

For $y(0.5)$, $0.5 \in [0,1]$, so substitute $x=0.5$ in $f1(x)$, we get

$f1(0.5) = -5.0625$

Example.3.34 Calculate Cubic Splines

X	1	2	3	4
Y	0	1	0	1

Find y(1.5).

Solution Cubic spline formula is

$$\int_{x_0}^{x_n} y dx = \sum_{i=1}^{n} \int_{x_{i-1}}^{x_i} \frac{1}{6h} \left[(x_i - x)^3 M_{i-1} + (x - x_{i-1})^3 M_i \right]$$

$$+ \frac{x_i - x}{h} \left(y_{i-1} - \frac{h^2}{6} M_{i-1} \right) + \frac{1}{h} (x - x_{i-1}) \left[y_i - \frac{h^2}{6} M_i \right] dx$$

$$M_{i-1} + 4M_i + M_{i+1} = \frac{6}{h^2} (y_{i-1} - 2y_i + y_{i+1}), for\, i = 1, 2, \ldots n - 1$$

Here $h=1, n=3$, $M_0 = 0, M_3 = 0$

Substitute $i=1$ in equation (2)

$$M_0 + 4M_1 + M_2$$

$$= \frac{6}{h^2} (y_0 - 2y_1 + y_2)$$

$$\Rightarrow 0 + 4M_1 + M_2 = 61 \cdot (0 - 2 \cdot 1 + 0)$$

$$\Rightarrow 4M_1 + M_2 = -12$$

Substitute $i=2$ in equation (2)

$$M_1 + 4M_2 + M_3$$

$$= \frac{6}{h^2}(y_1 - 2y_2 + y_3)$$

$$\Rightarrow M_1 + 4M_2 + 0 = 61 \cdot (1 - 2 \cdot 0 + 1)$$

$$\Rightarrow M_1 + 4M_2 = 12$$

Solving these 2 equations using elimination method

Substitute $i=1$ in equation (1), we get cubic spline in $1st$ interval $[x0,x1]=[1,2]$

$$f_1(x) = (x_1 - x)36hM_0 + (x - x_0)36hM_1$$

$$+(x_1 - x)h(y_0 - \frac{h^2}{6}M_0) + (x - x_0)h(y_1 - \frac{h^2}{6}M_1)f_1(x)$$

$$f_1(x) = (2 - x)36 \cdot 0 + (x - 1)36 \cdot -4$$

$$+(2 - x)1(0 - 16 \cdot 0) + (x - 1)1(1 - 16 \cdot -4)f1(x)$$

$$= 13(-2x_3 + 6x_2 - x - 3), \, for \, 1 \leq x \leq 2$$

Substitute $i=2$ in equation (1), we get cubic spline in $2nd$ interval $[x1,x2]=[2,3]$

$$f_2(x) = (x_2 - x)\frac{36}{h}M_1 + (x - x_1)\frac{36}{h}M_2$$

$$+(x_2 - x)h(y_1 - \frac{h^2}{6}M_1) + (x - x_1)h(y_2 - \frac{h^2}{6}M_2)f_2(x)$$

$$f2(x) = (3 - x)36 \cdot -4 + (x - 2)36 \cdot 4 + (3 - x)1(1 - 16 \cdot -4)$$

$$+(x - 2)1(0 - 16 \cdot 4)f2(x)$$

$$= 13(4x_3 - 30x_2 + 71x - 51), \, for \, 2 \leq x \leq 3$$

Substitute $i=3$ in equation (1), we get cubic spline in $3rd$ interval $[x_2,x_3]=[3,4]$

$$f_3(x) = (x_3 - x)\frac{36}{h}M_2 + (x - x_2)\frac{36}{h}M_3 + (x_3 - x)h(y_2 - \frac{h^2}{6}M_2)$$

$$+(x - x_2)h(y_3 - \frac{h^2}{6}M_3)f_3(x)$$

$$f_3(x) = (4 - x)36 \cdot 4 + (x - 3)36 \cdot 0 + (4 - x)1(0 - 16 \cdot 4)$$

$$+(x - 3)1(1 - 16 \cdot 0)f3(x)$$

$$= 13(-2x_3 + 24x_2 - 91x + 111), \, for \, 3 \le x \le 4$$

For $y(1.5)$, $1.5 \in [1,2]$, so substitute $x=1.5$ in $f_1(x)$, we get
$f_1(1.5) = 0.75$

Example.3.34 Calculate Cubic Splines y(1.5)

X	1	2	3	4	5
Y	0	1	0	1	0

Solution Cubic spline formula is

$$\int_{x_0}^{x_n} y\,dx = \sum_{i=1}^{n} \int_{x_{i-1}}^{x_i} \frac{1}{6h}\left[(x_i - x)^3 M_{i-1} + (x - x_{i-1})^3 M_i\right]$$

$$+\frac{x_i - x}{h}\left(y_{i-1} - \frac{h^2}{6}M_{i-1}\right) + \frac{1}{h}(x - x_{i-1})\left[y_i - \frac{h^2}{6}M_i\right]dx$$

$$M_{i-1} + 4M_i + M_{i+1} = \frac{6}{h^2}(y_{i-1} - 2y_i + y_{i+1}), \, for \, i = 1,2,...n-1$$

Here $h=1, n=4$

$$M_0 = 0, M_4 = 0$$

Substitute $i=1$ inequation (2)

$$M_0 + 4M_1 + M_2$$
$$= \frac{6}{h^2}(y_0 - 2y_1 + y_2)$$
$$\Rightarrow 0 + 4M_1 + M_2$$
$$= 61 \cdot (0 - 2 \cdot 1 + 0)$$
$$\Rightarrow 4M_1 + M_2 = -12$$

Substitute $i=2$ in equation (2)

$$M_1 + 4M_2 + M_3$$
$$= \frac{6}{h^2}(y_1 - 2y_2 + y_3)$$
$$\Rightarrow M_1 + 4M_2 + M_3$$
$$= 61 \cdot (1 - 2 \cdot 0 + 1)$$
$$\Rightarrow M_1 + 4M_2 + M_3 = 12$$

Substitute $i=3$ inequation (2)

$$M_2 + 4M_3 + M_4$$
$$= \frac{6}{h^2}(y_2 - 2y_3 + y_4)$$
$$\Rightarrow M_2 + 4M_3 + 0 = 61 \cdot (0 - 2 \cdot 1 + 0)$$
$$\Rightarrow M_2 + 4M_3 = -12$$

Solving these 3 equations using elimination method
Substitute $i=1$ in equation (1), we get cubic spline
in 1st interval $[x_0, x_1]=[1,2]$

$$f_1(x) = (x_1 - x)36hM_0 + (x - x_0)36hM_1$$

$$+(x_1 - x)h(y_0 - \frac{h^2}{6}M_0) + (x - x_0)h(y_1 - \frac{h^2}{6}M_1)f_1(x)$$

$$= (2 - x)36 \cdot 0 + (x - 1)36 - 4.2857 + (2 - x)1(0 - 16 \cdot 0)$$

$$+(x - 1)1(1 - 16 \cdot - 4.2857)f_1(x)$$

$$= -0.7143x_3 + 2.1428x_2 - 0.4286x - 1, \ for \ 1 \le x \le 2$$

Substitute $i=2$ in equation (1), we get cubic spline
in 2nd interval $[x_1, x_2] = [2,3]$

$$f_2(x) = (x_2 - x)\frac{36}{h}M_1 + (x - x_1)\frac{36}{h}M_2$$

$$+(x_2 - x)h(y_1 - \frac{h^2}{6}M_1) + (x - x_1)h(y_2 - \frac{h^2}{6}M_2)f_2(x)$$

$$= (3 - x)36 - 4.2857 + (x - 2)36 \cdot 5.1429$$

$$+(3 - x)1(1 - 16 \cdot - 4.2857)$$

$$+(x - 2)1(0 - 16 \cdot 5.1429)$$

$$f_2(x) = 1.5714x_3 - 11.5714x_2 + 27x - 19.2857$$

$$for \ 2 \le x \le 3$$

Substitute $i=3$ in equation (1), we get cubic spline
in 3rd interval $[x_2, x_3] = [3,4]$

$$f_3(x) = (x_3 - x)\frac{36}{h}M_2 + (x - x_2)\frac{36}{h}M_3$$

$$+(x_3 - x)h(y_2 - \frac{h^2}{6}M_2) + (x - x_2)h(y_3 - \frac{h^2}{6}M_3)f_3(x)$$

$$f_3(x) = (4 - x)36 \cdot 5.1429 + (x - 3)36 - 4.2857$$

$$+(4 - x)1(0 - 16 \cdot 5.1429) + (x - 3)1(1 - 16 \cdot - 4.2857)$$

$$f_3(x) = -1.5714x_3 + 16.7144x_2 - 57.8574x + 65.5718,$$

$$for \ 3 \le x \le 4$$

Substitute $i=4$ in equation (1), we get cubic spline in $4th$ interval $[x_3,x_4]=[4,5]$

$$f_4(x) = (x_4 - x)\frac{36}{h}M_3 + (x - x_3)\frac{36}{h}M_4$$

$$+(x_4 - x)h(y_3 - \frac{h^2}{6}M_3) + (x - x_3)h(y_4 - \frac{h^2}{6}M_4)$$

$$f_4(x) = (5 - x)36 - 4.2857 + (x - 4)36 \cdot 0$$

$$+(5 - x)1(1 - 16 \cdot -4.2857) + (x - 4)1(0 - 16 \cdot 0)$$

$$f_4(x) = 0.7143x_3 - 10.7142x_2 + 51.857x - 80.714,$$
$$for\ 4 \le x \le 5$$

For $y(1.5)$, $1.5 \in [1,2]$, so substitute $x=1.5$ in $f_1(x)$, we get

$$f1(1.5)=0.7679$$

Example.3.35 Calculate Cubic Splines

X	1	2	3	4	5
Y	1	2	5	11	20

Find y(1.5).

Solution Cubic spline formula is

$$\int_{x_0}^{x_n} ydx = \sum_{i=1}^{n} \int_{x_{i-1}}^{x_i} \frac{1}{6h}\left[(x_i - x)^3 M_{i-1} + (x - x_{i-1})^3 M_i\right]$$

$$+\frac{x_i - x}{h}\left(y_{i-1} - \frac{h^2}{6}M_{i-1}\right) + \frac{1}{h}(x - x_{i-1})\left[y_i - \frac{h^2}{6}M_i\right]dx$$

$$M_{i-1} + 4M_i + M_{i+1} = \frac{6}{h^2}\left(y_{i-1} - 2y_i + y_{i+1}\right),$$

$$\text{for } i = 1, 2, \ldots n-1$$

Here $h=1, n=4$

$$M_0 = 0, M_4 = 0$$

Substitute $i=1$ in

$$M_0 + 4M_1 + M_2 = \frac{6}{h^2}(y_0 - 2y_1 + y_2)$$

$$\Rightarrow 0 + 4M_1 + M_2 = 61 \cdot (1 - 2 \cdot 2 + 5)$$

$$\Rightarrow 4M_1 + M_2 = 12$$

Substitute $i=2$ in

$$M_1 + 4M_2 + M_3$$

$$= \frac{6}{h^2}(y_1 - 2y_2 + y_3)$$

$$\Rightarrow M_1 + 4M_2 + M_3 = 61 \cdot (2 - 2 \cdot 5 + 11)$$

$$\Rightarrow M_1 + 4M_2 + M_3 = 18$$

Substitute $i=3$ in
$$M_2 + 4M_3 + M_4$$

$$= \frac{6}{h^2}(y_2 - 2y_3 + y_4)$$

$$\Rightarrow M_2 + 4M_3 + 0 = 61 \cdot (5 - 2 \cdot 11 + 20)$$

$$\Rightarrow M_2 + 4M_3 = 18$$

Solving these 3 equations using elimination method
Substitute $i=1$ in equation (1), we get cubic spline
in 1st interval $[x_0, x_1] = [1, 2]$

$$f_1(x) = (x_1 - x)\frac{36}{h}M_0 + (x - x_0)\frac{36}{h}M_1$$

$$+(x_1 - x)h(y_0 - \frac{h^2}{6}M_0) + (x - x_0)h(y_1 - \frac{h^2}{6}M_1)$$

$$f_1(x) = (2 - x)36 \cdot 0 + (x - 1)36 \cdot 2.25$$

$$+(2 - x)1(1 - 16 \cdot 0) + (x - 1)1(2 - 16 \cdot 2.25)f1(x)$$

$$= 0.375x_3 - 1.125x_2 + 1.75x, \; for \, 1 \leq x \leq 2$$

Substitute $i=2$ in equation (1), we get cubic spline in 2nd interval $[x_1, x_2]=[2,3]$

$$f_2(x) = (x_2 - x)\frac{36}{h}M_1 + (x - x_1)\frac{36}{h}M_2 + (x_2 - x)h(y_1 - \frac{h^2}{6}M_1)$$

$$+(x - x_1)h(y_2 - \frac{h^2}{6}M_2)$$

$$f_2(x) = (3 - x)36 \cdot 2.25 + (x - 2)36 \cdot 3$$

$$+(3 - x)1(2 - 16 \cdot 2.25) + (x - 2)1(5 - 16 \cdot 3)f2(x)$$

$$= 0.125x_3 + 0.375x_2 - 1.25x + 2, \; for \, 2 \leq x \leq 3$$

Substitute $i=3$ in equation (1), we get cubic spline in 3rd interval $[x_2, x_3]=[3,4]$

$$f_3(x) = (x_3 - x)\frac{36}{h}M_2 + (x - x_2)\frac{36}{h}M_3 + (x_3 - x)h(y_2 - \frac{h^2}{6}M_2)$$

$$+(x - x_2)h(y_3 - \frac{h^2}{6}M_3)$$

$$f3(x) = (4 - x)36 \cdot 3 + (x - 3)36 \cdot 3.75$$

$$+(4 - x)1(5 - 16 \cdot 3) + (x - 3)1(11 - 16 \cdot 3.75)f3(x)$$

$$= 0.125x_3 + 0.375x_2 - 1.25x + 2, \; for \, 3 \leq x \leq 4$$

Substitute $i=4$ in equation (1), we get cubic spline in 4th interval $[x_3, x_4]=[4,5]$

$$f_4(x) = (x_4 - x)\frac{36}{h}M_3 + (x - x_3)\frac{36}{h}M_4$$

$$+(x_4 - x)h(y_3 - \frac{h^2}{6}M_3) + (x - x_3)h(y_4 - \frac{h^2}{6}M_4)$$

$$f_4(x) = (5-x)36\cdot3.75 + (x-4)36\cdot0$$

$$+(5-x)1(11-16\cdot3.75) + (x-4)1(20-16\cdot0)f4(x)$$

$$= -0.625x_3 + 9.375x_2 - 37.25x + 50, \text{ for } 4 \le x \le 5$$

For $y(1.5)$, $1.5 \in [1,2]$, so substitute $x=1.5$ in $f_1(x)$, we get

$f_1(1.5) = 1.3594$.

UNIT-4

Numerical differentiation and integration

4.1. Numerical differentiation

It is the process of calculating the value of the derivative of a function at some assigned value of x from the given set of values (x_i, y_i). To compute $\dfrac{dy}{dx}$, we first replace the exact relation $y = f(x)$ by the best interpolating polynomial $y = \phi(x)$ and then differentiate the latter as many times as we desire. The choice of the interpolation formula to be used, will depend on the assigned value of x at which $\dfrac{dy}{dx}$ is desired. If the values of x are equispaced and $\dfrac{dy}{dx}$ is required near the beginning of the table, we use Newton's forward formula. If it is required near the end of the table, we use Newton 's backward formula. For values near the middle of the table, $\dfrac{dy}{dx}$ is calculated by means of Stirling's or Bessel's formula or Newton's divided difference formula to represent the function. Hence corresponding to each of the interpolation formulae, we can derive a formula for finding the derivative.

It is a technique of numerical analysis to produce an estimate of the derivative of a mathematical function subroutine using values of the function. If the definition of the function is given explicitly, then its derivative can be found usually. For tabulated functions, i.e., if the values of the function for a discrete set of values are given, then the derivatives can be calculated by numerical differentiation. In some cases, the functions may be highly complex, so we seek the help of numerical differentiation to estimate its derivatives or integrals.

Observation. While using these formulae, it must be observed that the table of values defines the function at these points only and does not completely define the function and the function may not be differentiable at all. As such, the process of numerical differentiation should be used only if the tabulated values are such that the differences of some order are constants. Otherwise, errors are bound to creep in which go on increasing as derivatives of higher order are found. This is due to the fact that the difference between $f(x)$ and the approximating polynomial $\phi(x)$ may be small at the data points but $f^1(x) - \phi^1(x)$ may be large.

Numerical differentiation is carried out in 2 cases.

1. If the function is unknown and if it is given in tabular form.
2. If the function is highly complex.
   In numerical differentiation, we will replace the function
   $y = f(x)$ on [a, b]

With an interpolating polynomial $p(x)$. It is less exact than interpolation. When the values of argument x for $y = f(x)$ are equally spaced and $\dfrac{dy}{dx}$ is required at the beginning of the table, then Newton's forward interpolation formula is used to represent the function. When $\dfrac{dy}{dx}$ is required at a point nearer to the end of the table, then Newton's backward interpolation formula is used. When the derivative is to be found at some point lying near the middle of the table, then central difference interpolation formula is used. If the values of the argument x are not equally spaced, then Newton's divided difference formula or Lagrange's interpolation formula is used to approximate the function $y = f(x)$.

4.1.1. Derivatives using Bessel's central difference formula.

Bessel's formula

$$y_p = y_0 + p\Delta y_0 + \frac{p(p-1)}{2}\frac{\Delta^2 y_{-1} + \Delta^2 y_0}{2}$$

$$+ \left(p - \frac{1}{2}\right)\frac{p(p-1)}{\angle 3}\Delta^3 y_{-1} + ...,$$

$$\text{where } p = \frac{x - x_0}{h}, \therefore \frac{dp}{dx} = \frac{1}{h}$$

$$\text{Now,} \frac{dy}{dx} = \frac{dy}{dp}.\frac{dp}{dx} = \frac{1}{h}\left[\begin{array}{c} \Delta y_0 + \frac{2p-1}{\angle 2}\frac{\Delta^2 y_{-1} + \Delta^2 y_0}{2} \\ + \frac{3p^2 - 2p + \frac{1}{2}}{\angle 3}\Delta^3 y_{-1} \\ + ... \end{array}\right]$$

At $x = x_0, p = 0$.

Hence putting p=0, we get

$$\left(\frac{dy}{dx}\right)_{x_0} = \frac{1}{h}\left[\begin{array}{l}\Delta y_0 - \frac{1}{2}\left(\frac{\Delta^2 y_{-1} + \Delta^2 y_0}{2}\right) + \frac{1}{12}\Delta^3 y_{-1} \\ + \frac{1}{12}\left(\frac{\Delta^4 y_{-2} + \Delta^4 y_{-1}}{2}\right) \\ + \dots \end{array}\right]$$

$$\left(\frac{d^2 y}{dx^2}\right)_{x_0} = \frac{1}{h^2}\left[\begin{array}{l}\left(\frac{\Delta^2 y_{-1} + \Delta^2 y_0}{2}\right) - \frac{1}{2}\Delta^3 y_{-1} \\ - \frac{1}{12}\left(\frac{\Delta^4 y_{-2} + \Delta^4 y_{-1}}{2}\right) \\ + \dots \end{array}\right]$$

4.1.2. First derivative of newton's forward difference formula

Newton's forward difference formula is

$$y = y_0 + p\Delta y_0 + \frac{p(p-1)}{2}\Delta^2 y_0$$
$$+ \frac{p(p-1)(p-2)}{\angle 3}\Delta^3 y_0 + \dots\dots$$
$$p = \frac{x - x_0}{h}$$
$$\Rightarrow \frac{dp}{dx} = \frac{1}{h}$$

$$Now \frac{dy}{dx} = 0 + \Delta y_0 + \frac{2p-1}{2}\Delta^2 y_0$$

$$+ \frac{3p^2 - 6p + 2}{6}\Delta^3 y_0 + ...$$

$$h.\frac{dy}{dx} = \Delta y_0 + \frac{2p-1}{2}\Delta^2 y_0 + \frac{3p^2 - 6p + 2}{6}\Delta^3 y_0$$

$$+.....$$

$$\frac{dy}{dx} = \frac{1}{h}\left[\begin{array}{c}\Delta y_0 + \frac{2p-1}{2}\Delta^2 y_0 + \frac{3p^2 - 6p + 2}{6}\Delta^3 y_0 \\ +.....\end{array}\right]$$

$$At x = x_0 \Rightarrow p = 0$$

$$\Rightarrow \left(\frac{dy}{dx}\right)_{x=x_0} = \frac{1}{h}\left[\Delta y_0 + \frac{-1}{2}\Delta^2 y_0 + \frac{2}{6}\Delta^3 y_0 + ...\right]$$

$$\Rightarrow \left(\frac{dy}{dx}\right)_{x=x_0} = \frac{1}{h}\left[\begin{array}{c}\Delta y_0 - \frac{1}{2}\Delta^2 y_0 + \frac{1}{3}\Delta^3 y_0 - \frac{1}{4}\Delta^4 y_0 \\ +.....\end{array}\right]$$

4.13. Second derivative of Newton's forward difference formula

$$\frac{d^2 y}{dx^2} = \frac{d}{dx}\left(\frac{dy}{dx}\right) = \frac{d}{dp}\left(\frac{dy}{dx}\right)\frac{dp}{dx} = \frac{d}{dp}\left(\frac{dy}{dx}\right)\frac{1}{h}$$

$$\Rightarrow \frac{d^2 y}{dx^2} = \frac{d}{dp}\frac{1}{h}\left[\frac{1}{h}\left[\Delta y_0 + \frac{2p-1}{2}\Delta^2 y_0 + \frac{3p^2 - 6p + 2}{6}\Delta^3 y_0 + ...\right]\right]$$

$$= \frac{1}{h^2}\left[\frac{2}{2}\Delta^2 y_0 + \frac{6p-6}{6}\Delta^3 y_0 + ...\right]$$

At $x = x_0, p = 0$

$$\left(\frac{d^2 y}{dx^2}\right)_{x=x_0} = \frac{1}{h^2}\left[\Delta^2 y_0 + \frac{0-6}{6}\Delta^3 y_0 + \frac{0+11}{12}\Delta^4 y_0 + \ldots\ldots\right]$$

$$\left(\frac{d^2 y}{dx^2}\right)_{x=x_0} = \frac{1}{h^2}\left[\Delta^2 y_0 - \Delta^3 y_0 + \frac{11}{12}\Delta^4 y_0 + \ldots\ldots\right]$$

similarly,

$$\left(\frac{d^3 y}{dx^3}\right)_{x=x_0} = \frac{1}{h^3}\left[\Delta^3 y_0 - \frac{3}{2}\Delta^4 y_0 + \ldots\ldots\right]$$

We can also find the above formulae in the following way,

We know that

$$E = e^{hD}$$

$$\Rightarrow 1 + \Delta = e^{hD}$$

$$\Rightarrow hD = \log(1+\Delta)$$

$$\Rightarrow hD = \Delta - \frac{\Delta^2}{2} + \frac{\Delta^3}{3} - \frac{\Delta^4}{4} + \ldots\ldots$$

$$\left[\because \log(1+x) = x - \frac{x^2}{2} + \frac{x^3}{3} - \frac{x^4}{4} + \ldots\ldots\right]$$

$$\Rightarrow D = \frac{1}{h}\left[\Delta - \frac{\Delta^2}{2} + \frac{\Delta^3}{3} - \frac{\Delta^4}{4} + \ldots\ldots\right]$$

$$\Rightarrow D^2 = \frac{1}{h^2}\left[\Delta - \frac{\Delta^2}{2} + \frac{\Delta^3}{3} - \frac{\Delta^4}{4} + \ldots\ldots\right]^2$$

$$\Rightarrow D^2 = \frac{1}{h^2}\left[\Delta^2 - \Delta^3 + \frac{11}{12}\Delta^4 - \frac{5}{6}\Delta^5 + \ldots\ldots\right]$$

Then $Dy_0 = \dfrac{1}{h}\left[\Delta y_0 - \dfrac{\Delta^2 y_0}{2} + \dfrac{\Delta^3 y_0}{3} - \dfrac{\Delta^4 y_0}{4} + \ldots\ldots\right]$

$D^2 y_0 = \dfrac{1}{h^2}\left[\Delta^2 y_0 - \Delta^3 y_0 + \dfrac{11}{3}\dfrac{\Delta^4 y_0}{4} - \dfrac{5}{6}\Delta^5 y_0 + \ldots\ldots\ldots\right]$

4.1.4. First derivative of newton's backward difference formula

Newton's backward difference formula is

$$y = y_n + p\nabla y_n + \dfrac{p(p+1)}{2}\nabla^2 y_0 + \dfrac{p(p+1)(p+2)}{\angle 3}\nabla^3 y_0$$

$$+\ldots\ldots$$

$$p = \dfrac{x - x_n}{h}$$

$$\Rightarrow \dfrac{dp}{dx} = \dfrac{1}{h}$$

$$\left(\dfrac{dy}{dx}\right) = \left(\dfrac{dy}{dp}\right)\dfrac{dp}{dx} = \left(\dfrac{dy}{dp}\right)\dfrac{1}{h} \Rightarrow h\dfrac{dy}{dx} = \dfrac{dy}{dp}$$

$Diff.(\ )w.\mathrm{r.t.p.}$

$Now\dfrac{dy}{dx} = 0 + \nabla y_n + \dfrac{2p+1}{2}\nabla^2 y_n + \dfrac{3p^2 + 6p + 2}{6}\nabla^3 y_n + \ldots$

$h.\dfrac{dy}{dx} = \nabla y_n + \dfrac{2p+1}{2}\nabla^2 y_n + \dfrac{3p^2 + 6p + 2}{6}\nabla^3 y_n + \ldots$

190

$$\frac{dy}{dx} = \frac{1}{h}\left[\nabla y_0 + \frac{2p+1}{2}\nabla^2 y_n + \frac{3p^2+6p+2}{6}\nabla^3 y_n + ...\right]$$

$At\, x = x_n \Rightarrow p = 0$

$$\Rightarrow \left(\frac{dy}{dx}\right)_{x=x_n} = \frac{1}{h}\left[\nabla y_n + \frac{1}{2}\nabla^2 y_n + \frac{2}{6}\nabla^3 y_n + ...\right]$$

$$\Rightarrow \left(\frac{dy}{dx}\right)_{x=x_0} = \frac{1}{h}\left[\nabla y_n + \frac{1}{2}\nabla^2 y_n + \frac{2}{6}\nabla^3 y_n + \frac{1}{4}\nabla^4 y_n.....\right]$$

4.1.5. Second derivative of Newton's backward difference formula

$$\frac{d^2 y}{dx^2} = \frac{d}{dx}\left(\frac{dy}{dx}\right) = \frac{d}{dp}\left(\frac{dy}{dx}\right)\frac{dp}{dx} = \frac{d}{dp}\left(\frac{dy}{dx}\right)\frac{1}{h}$$

$$\Rightarrow \frac{d^2 y}{dx^2} = \frac{d}{dp}\frac{1}{h}\left[\frac{1}{h}\left[\nabla y_n + \frac{2p+1}{2}\nabla^2 y_n + \frac{3p^2+6p+2}{6}\nabla^3 y_n + ...\right]\right]$$

$$= \frac{1}{h^2}\left[\frac{2}{2}\nabla^2 y_n + \frac{6p+6}{6}\nabla^3 y_n + ...\right]$$

$At\, x = x_n, p = 0$

$$\left(\frac{d^2 y}{dx^2}\right)_{x=x_n} = \frac{1}{h^2}\left[\nabla^2 y_n + \nabla^3 y_n + ...\right]$$

$$\left(\frac{d^2 y}{dx^2}\right)_{x=x_n} = \frac{1}{h^2}\left[\nabla^2 y_n + \nabla^3 y_n + \frac{11}{12}\nabla^4 y_n + \frac{5}{6}\nabla^5 y_n + ...\right]$$

similarly,

$$\left(\frac{d^3 y}{dx^3}\right)_{x=x_n} = \frac{1}{h^3}\left[\nabla^3 y_n - \frac{3}{2}\nabla^4 y_n + ...\right]$$

We can also find the above formulae in the following way,

We know that

$$E = e^{hD}$$

$$\Rightarrow (1-\nabla)^{-1} = e^{hD}$$

$$\Rightarrow hD = \log(1-\nabla)^{-1}$$

$$\Rightarrow hD = \nabla + \frac{\nabla^2}{2} + \frac{\nabla^3}{3} + \frac{\nabla^4}{4} + \ldots$$

$$\left[\because \log(1-x) = x + \frac{x^2}{2} + \frac{x^3}{3} + \frac{x^4}{4} + \ldots\right]$$

$$\Rightarrow D = \frac{1}{h}\left[\nabla + \frac{\nabla^2}{2} + \frac{\nabla^3}{3} + \frac{\nabla^4}{4} + \ldots\right]$$

$$\Rightarrow D^2 = \frac{1}{h^2}\left[\nabla + \frac{\nabla^2}{2} + \frac{\nabla^3}{3} + \frac{\nabla^4}{4} + \ldots\right]^2$$

$$\Rightarrow D^2 = \frac{1}{h^2}\left[\nabla^2 + \nabla^3 + \frac{11}{12}\nabla^4 + \frac{5}{6}\nabla^5 + \ldots\right]$$

$$\left(\frac{d^3 y}{dx^3}\right)_{x=x_n} = \frac{1}{h^3}\left(\nabla^3 y_n + \frac{3}{2}\nabla^4 y_n\right)$$

4.2. Maximum and minimum values of a tabulated function

Let $y = f(x)$ be a function. Let $y_0, y_1, y_2, \ldots y_n$ be the values of y for $x = x_0, x_1, x_2, \ldots x_n$. Now we find the values of x at which the function is maximum or minimum.

We know that the maximum and minimum values of a function can be found by equating the first derivative to zero then we find the x value by solving that equation.

Let $x_0, x_1, x_2, \ldots x_n$ are equally spaced with difference h.

Consider Newton's forward difference formula

$$y = y_0 + p\Delta y_0 + \frac{p(p-1)}{2}\Delta^2 y_0 + \frac{p(p-1)(p-2)}{6}\Delta^3 y_0 + \ldots\ldots$$

Differentiate w.r.t p

$$\frac{dy}{dp} = 0 + \Delta y_0 + \frac{2p-1}{2}\Delta^2 y_0 + \frac{3p^2-6p+2}{6}\Delta^3 y_0 + \ldots\ldots$$

For maxima and minima $\frac{dy}{dp} = 0$, on R. H. S, we will take it to 3 differences

$$\Delta y_0 + \frac{2p-1}{2}\Delta^2 y_0 + \frac{3p^2-6p+2}{6}\Delta^3 y_0 = 0$$

$$p^2\left(\frac{1}{2}\Delta^3 y_0\right) + p\left(\Delta^2 y_0 - \Delta^3 y_0\right) + \left(\Delta y_0 - \frac{1}{2}\Delta^2 y_0 + \frac{1}{3}\Delta^3 y_0\right) = 0$$

We put the values of $\Delta y_0, \Delta^2 y_0, \Delta^3 y_0$ from the table and

$$p = \frac{x-x_0}{h} \Rightarrow x = x_0 + ph$$

We can find the values of x for which y is maximum or minimum.

Example.4.1 From the following table, find x for which y is maximum and find this value of y.

X	1.2	1.3	1.4	1.5	1.6
Y	0.9320	0.9636	0.9855	0.9975	0.9996

Solution　　　Given, $x_0 = 1.2$

We have

$$\frac{dy}{dp} = \Delta y_0 + \frac{2p-1}{2}\Delta^2 y_0 + \frac{3p^2 - 6p + 2}{6}\Delta^3 y_0 + ...$$

X	Y	Δ	Δ^2	Δ^3
1.2	0.9320			
1.3	0.9636	0.0316		
1.4	0.9855	0.0219	-0.0097	
1.5	0.9975	0.0120	-0.0099	-0.0002
1.6	0.9996	0.0021	-0.0099	0

$$\frac{dy}{dp} = 0.0316 + \frac{2p-1}{2}(-0.0097) = -0.0316$$

$$\Rightarrow 2p - 1 = \frac{0.0632}{0.0097}$$

$$\Rightarrow p = 3.8$$

$$x = x_0 + ph \Rightarrow x = 1.2 + (3.8)(0.1) = 1.58$$

Since x is nearer to 1.6 at the end of the table, Newton's backward difference formula is

$$y = y_n + p\nabla y_n + \frac{p(p+1)}{2}\nabla^2 y_n + \ldots..$$

$$x = 1.58, x_n = 1.6, p = \frac{x - x_n}{h} = \frac{1.58 - 1.6}{0.1} = -0.2$$

$$y = 0.9996 + (-0.2)(0.0021) + \frac{(-0.2)(-0.2+1)}{2}(-0.0099)$$

$$y = 0.9996 - 0.0004 + 0.0008$$

$$y = 1$$

Example.4.2

Given that X	y
1.0	7.989
1.1	8.403
1.2	8.781
1.3	9.129
1.4	9.451
1.5	9.750
1.6	10.031

Find $\dfrac{dy}{dx}$ and $\dfrac{d^2 y}{dx^2}$ at x=1.1, x=1.6.

Solution To find x=1.1.

X	Y	Δ	Δ^2	Δ^3	Δ^4	Δ^5	Δ^6
1.0	7.989						
1.1	8.403	0.414					
1.2	8.781	0.378	- 0.036				
1.3	9.129	0.348	-	0.006			

195

1.4	9.451	0.322	0.030				
			-0.026	0.004	-0.002		
1.5	9.750	0.299	-0.023	0.003	-0.001	0.001	
1.6	10.031	0.281	-0.018	0.005	0.002	0.003	0.002

We have

$$\left(\frac{dy}{dx}\right)_{x_0} = \frac{1}{h}\left[\begin{array}{l}\Delta y_0 - \frac{1}{2}\Delta^2 y_0 + \frac{1}{3}\Delta^3 y_0 + \frac{1}{4}\Delta^4 y_0 + \frac{1}{5}\Delta^5 y_0 \\ -\frac{1}{6}\Delta^6 y_0 + ...\end{array}\right]$$

$$\left(\frac{d^2 y}{dx^2}\right)_{x_0} = \frac{1}{h^2}\left[\begin{array}{l}\Delta^2 y_0 - \Delta^3 y_0 + \frac{11}{12}\Delta^4 y_0 - \frac{5}{6}\Delta^5 y_0 \\ +\frac{137}{180}\Delta^6 y_0 - .....\end{array}\right]$$

$h = 0.1, x_0 = 1.1, \Delta y_0 = 0.378, \Delta^2 y_0 = -0.03$ etc.

Substituting these values in i and ii, we get

$$\left(\frac{dy}{dx}\right)_{x_0} = \frac{1}{0.1}\left[\begin{array}{l}0.378 - \frac{1}{2}(-0.03) + \frac{1}{3}(0.004) \\ -\frac{1}{4}(-0.001) + \frac{1}{5}(0.003)\end{array}\right]$$

$$= 3.952$$

$$\left(\frac{d^2 y}{dx^2}\right) = \frac{1}{(0.1)^2}\left[\begin{array}{l}-0.03 - 0.004 \\ +\frac{11}{12}(-0.001) - \frac{5}{6}(0.003)\end{array}\right]$$

$$= -3.74$$

b. We use the above difference table and the backward difference operator ∇ instead of Δ, we get,

$$\left(\frac{dy}{dx}\right)_{x_0} = \frac{1}{h}\left[\nabla y_0 + \frac{1}{2}\nabla^2 y_n + \frac{1}{3}\nabla^3 y_n + \frac{1}{4}\nabla^4 y_n + \frac{1}{5}\nabla^5 y_n + \frac{1}{6}\nabla^6 y_n + ...\right]$$

$$\left(\frac{d^2 y}{dx^2}\right)_{x_0} = \frac{1}{h^2}\left[\nabla^2 y_n + \nabla^3 y_n + \frac{11}{12}\nabla^4 y_n + \frac{5}{6}\nabla^5 y_n + \frac{137}{180}\nabla^6 y_n + ...\right]$$

$h = 0.1, x_n = 1.6, \nabla y_n = 0.281, \nabla^2 y_n = -0.018$ etc.

Putting these values in i and ii, we get,

$$\left(\frac{dy}{dx}\right)_{1.0} = \frac{1}{0.1}\left[\begin{array}{l}0.281+\frac{1}{2}(-0.018)+\frac{1}{3}(0.05)+\frac{1}{4}(0.002) \\ +\frac{1}{5}(0.003)+\frac{1}{6}(0.002)\end{array}\right]$$

$$= 2.75$$

$$\left(\frac{d^2 y}{dx^2}\right)_{1.0} = \frac{1}{(0.1)^2}\left[\begin{array}{l}(-0.018)+(0.005)+\frac{11}{12}(0.002) \\ +\frac{5}{6}(0.003)+\frac{137}{180}(0.002)\end{array}\right]$$

$$= -0.715$$

Example.4.3

The following data gives the velocity of a particle for twenty seconds at an interval of five seconds. Find the initial acceleration using the entire data

Time t	0	5	10	15	20
Velocity v	0	3	14	69	228

Solution Let

t	v	Δv	$\Delta^2 v$	$\Delta^3 v$
0	0			
5	3	36	8	24
10	14	60	44	
15	69		104	
20	228			

An initial acceleration i.e., $\left(\dfrac{dv}{dt}\right)$ at t=0 is required, we use

Newton's forward formula,

$$\left(\frac{dv}{dt}\right)_{t=0} = \frac{1}{h}\left[\Delta v_0 - \frac{1}{2}\Delta^2 v_0 + \frac{1}{3}\Delta^3 v_0 - \frac{1}{4}\Delta^4 v_0 + ...\right]$$

$$\therefore \left(\frac{dv}{dt}\right)_{t=0} = \frac{1}{5}\left[3 - 8 + \frac{1}{3}(36) - \frac{1}{4}(24)\right]$$

$$= \frac{1}{5}(3 - 4 + 12 - 6) = 1$$

Hence the initial acceleration is 1 m/sec^2.

Example. 4.4

Find the value of cos 1.74 from the following table.

X	1.7	1.74	1.78	1.82	1.86
Sinx	0.9916	0.9857	0.9781	0.9691	0.9584

Solution Let $y = f(x) = \sin x \Rightarrow f'(x) = \cos x$

The difference table is

198

x	y	Δy	$\Delta^2 y$	$\Delta^3 y$	$\Delta^4 y$
1.7	0.9916				
1.74	0.9857	-0.0059			
1.78	0.9781	-0.0070	-0.0017		
1.82	0.9691	-0.0090	-0.0014	0.003	
1.84	0.9584	-0.0107	-0.0017	-0.003	-0.0006

Since we require $f'(1.74)$, we use Newton's forward formula

$$\frac{dy}{dx} = \frac{1}{h}\left[\Delta y_0 - \frac{1}{2}\Delta^2 y_0 + \frac{1}{3}\Delta^3 y_0 - \frac{1}{4}\Delta^4 y_0 + ...\right]$$

Here h=0.04, $x_0 = 1.7, \Delta y_0 = -0.0059, \Delta^2 y_0 = -0.0017 etc$

Substituting these values in I, we get

$$\left(\frac{dy}{dx}\right)_{1.74} = \frac{1}{0.04}\left[\begin{array}{c} 0.0059 - \frac{1}{2}(-0.0017) \\ + \frac{1}{3}(0.003) - \frac{1}{4}(-0.006) \end{array}\right]$$

$$= \frac{1}{0.04}(0.007) = 0.175$$

$$\cos(1.74) = 0.175$$

Example.4.5

A slider in a machine moves along a fixed straight rod. Its distance x cm. along the rod is given below for various values of the time t seconds. Find the velocity of the slider and its acceleration when t=0.3 second.

t	0	0.1	0.2	0.3	0.4	0.5	0.6
y	30.13	31.62	32.87	33.64	33.95	33.81	33.24

Solution The difference table is

t	y	Δy	$\Delta^2 y$	$\Delta^3 y$	$\Delta^4 y$	$\Delta^5 y$	$\Delta^6 y$
0	30.13						
0.1	31.62	1.49					
0.2	32.87	1.25	-0.24				
0.3	33.64	0.77	-0.48	-0.24			
0.4	33.95	0.31	-0.46	0.02	0.26		
0.5	33.81	-0.14	-0.45	0.01	-0.01	-0.27	
0.5	33.24	-0.57	-0.43	0.02	0.01	0.02	0.29

As the derivatives ere required near the middle of the table, we use Stirling's formulae.

$$\left(\frac{dx}{dt}\right)_{t_0} = \frac{1}{h}\left(\frac{\Delta x_0 + \Delta x_{-1}}{2}\right) - \frac{1}{6}\left(\frac{\Delta^3 x_{-1} + \Delta^3 x_{-2}}{2}\right) + \frac{1}{30}\left(\frac{\Delta^5 x_{-1} + \Delta^5 x_{-2}}{2}\right) + \dots.$$

$$\left(\frac{d^2 x}{dt^2}\right) = \frac{1}{h^2}\left[\Delta^2 x_{-1} - \frac{1}{12}\Delta^4 x_{-2} + \frac{1}{90}\Delta^6 x_{-3}\dots\dots\right]$$

Here h=0.1,

$t_0 = 0.3, \Delta x_0 = 0.31, \Delta x_{-1} = 0.77, \Delta^2 x_{-1} = -0.46$ *etc.*

Putting these values in i and ii, we get,

$$\left(\frac{dt}{dx}\right)_{0.3} = \frac{1}{0.1}\left[\frac{0.31+0.77}{2} - \frac{1}{6}\left(\frac{0.01+0.02}{2}\right) + \frac{1}{30}\left(\frac{0.02-0.27}{2}\right) - ...\right]$$
$$= 5.33$$

$$\left(\frac{d^2y}{dx^2}\right)_{0.3} = \frac{1}{0.01}\left[-0.46 - \frac{1}{12}(-0.01) + \frac{1}{90}(0.29) - .....\right] = -45.6$$

Hence the required velocity is 5.33 cm/sec and acceleration is -45.6 cm/sec^2

Example.4.6 The elevation above a datum line of seven points of a road are given below.

X	0	300	600	900	1200	1500	1800
y	135	149	157	183	201	205	193

Find the gradient of the road at the middle point.

Solution Here h=300, $x_0 = 0, y_0 = 135$, we require the gradient dy/dx at x=900.

The difference table is

x	y	Δy	$\Delta^2 y$	$\Delta^3 y$	$\Delta^4 y$	$\Delta^5 y$
0	135					
300	149	14				
600	157	8	-6			
900	183	20	18	24		
1200	201	18	-8	-26	-50	
1500	205	4	-14	-6	20	70
1800	193	-12	-16	-2	4	-16

Using Stirling's formula for the first derivative, we get

$$y'(x_0) = \frac{1}{h}\left[\begin{array}{c}\left(\dfrac{\Delta y_0 + \Delta y_{-1}}{2}\right) - \dfrac{1}{6}\left(\dfrac{\Delta^3 y_{-1} + \Delta^3 y_{-2}}{2}\right) \\ + \dfrac{1}{30}\left(\dfrac{\Delta^5 y_{-2} + \Delta^5 y_{-3}}{2}\right)\end{array}\right]$$

$$= \frac{1}{300}\left[\frac{1}{2}(18+26) - \frac{1}{12}(-6-26) + \frac{1}{60}(-16+70)\right]$$

$$= \frac{1}{300}(22 + 2.666 + 0.9) = 0.085$$

Hence the gradient of the road at the middle point is 0.085.

Example.4.7 Using Bessel's formula, find $f'(7.5)$ from the following table.

X	7.47	7.48	7.49	7.50	7.51	7.52	7.53
Y	0.193	0.195	0.198	0.201	0.203	0.206	0.208

Solution Taking

$$x_0 = 7.50, h = 0.1, p = \frac{x - x_0}{h} = \frac{x - 7.50}{0.01}$$

The difference table is

X	P	Yp	Δ	Δ^2	Δ^3	Δ^4	Δ^5
7.47	-3	0.193					
0.002							
7.48	-2	0.195		0.001			
0.003		-0.001					
7.49	-1	0.198		0.000	0.000		
0.003		-0.001		0.003			
7.50	0	0.201		-0.001	0.003		-

202

						0.01
0.002		0.002		-0.007		
7.51	1	0.203		0.001	-0.004	
0.003		-0.002				
7.52	2	0.206		-0.001		
0.002						
7.53	3	0.208				

Using Bessel's formula for the first derivatives, we get,

$$\left(\frac{dy}{dx}\right)_{x_0} = \frac{1}{h}\begin{bmatrix} \Delta y_0 - \frac{1}{4}\left(\Delta^2 y_{-1} + \Delta^2 y_0\right) + \frac{1}{12}\Delta^2 y_{-1} \\ + \frac{1}{24}\left(\Delta^4 y_{-2} + \Delta^4 y_{-1}\right) - \frac{1}{120}\Delta^5 y_{-2} \\ - \frac{1}{240}\left(\Delta^6 y_{-2} + \Delta^6 y_{-2}\right) \end{bmatrix}$$

$$\left(\frac{dy}{dx}\right)_{7.5} = \frac{1}{0.01}\begin{bmatrix} 0.002 - \frac{1}{4}(-0.001 + 0.001) + \frac{1}{12}(0.002) + \\ \frac{1}{24}(-0.004 + 0.003) - \frac{1}{120}(-0.007) \\ - \frac{1}{240}(-0.010 + 0) \end{bmatrix}$$

$$= 0.2 + 0 + 0.01666 - 0.0416 + 0.00583 + 0.00416$$

$$= 0.223$$

Example.4.8 Find $f'(10)$ from the following data

X	3	5	11	24	34
Y	13	23	899	17315	35606

Solution As the values of x are not equi spaced, we shall use Newton's divided difference formula. The divided difference table is

X	Y	1st diff.	2nd diff.	3rd diff.	4th diff.
3	-13				
		15			
5	23		16		
		146		0.998	
11	899		39.96		0.0002
		1025		1.003	
27	17315		69.04		
		2613			
34	35606				

Fifth differences being zero, Newton's divided difference formula for the first derivative, we get,

$$f'(x) = f(x_0, x_1) + (2x - x_0 - x_1) f(x_0, x_1, x_2) +$$
$$\left[3x^2 - 2x(x_0 + x_1 + x_2) + x_0 x_1 + x_1 x_2 + x_2 x_0 \right]$$
$$*f(x_0, x_1, x_2, x_3) + \begin{bmatrix} 4x^3 - 3x^2(x_0 + x_1 + x_3 + x_3) \\ +2x(x_0 x_1 + x_1 x_2 + x_2 x_3 + x_3 x_6 + x_1 x_3 + x_0 x_2) \\ -(x_0 x_1 x_2 + x_1 x_2 x_3 + x_2 x_3 x_0 + x_0 x_1 x_3) \end{bmatrix}$$
$$f(x_0, x_1, x_2, x_3. x_4)$$

Putting
$$x_0 = 3, x_1 = 5, x_2 = 11, x_3 = 27, x = 10$$
$$f'(x) = 18 + 12 + 16 + 23 \times 0.998 - 426 - 0.0002$$
$$= 232.869$$

Example. 4.9 From the table below, for what value of x, y is minimum. Also find this value of y.

X	3	4	5	6	7	8
y	0.205	0.240	0.250	0.202	0.250	0.224

Solution The difference table is

X	Y	Δ	Δ^2	Δ^3
3	0.205			
		0.035		
4	0.240		-0.016	
		0.019		0.000
5	0.259		-0.016	
		0.003		0.001
6	0.262		-0.015	
		-0.012		0.001
7	0.250		-0.014	
		-0.026		
8	0.224			

Taking $x_0 = 3, y_0 = 0.205, \Delta y_0 = 0.035, \Delta^2 y_0 = -0.016$ and remaining is zero.

By using, Newton's forward difference formula

$$y = 0.205 + p(0.035) + \frac{p(p-1)}{2}(-0.016)$$

Differentiating it w.r.t. p, we have

$$\frac{dy}{dp} = 0.035 + \frac{29-1}{2}(-0.016)$$

For y to be minimum, dy/dp=0

$$\therefore 0.035 - 0.008(2p - 1) = 0$$

Which gives p=2.6875

$$\because x = x_0 + ph = 3 + 2.6875 = 5.6875$$

Hence y is minimum when x=5.6875

Putting p=2.6875, the minimum value of y=0.205-2.6875*1.6875* 0.016/2=0.2628.

Example.4.10 Find the maximum and minimum value of y from the following data.

X	-2	-1	0	1	2	3	4
Y	2	-0.25	0	-0.25	2	15.75	56

Solution The difference table is

X	Y	Δy	$\Delta^2 y$	$\Delta^3 y$	$\Delta^4 y$	$\Delta^5 y$
-2	2					
		-2.25				
-1	-0.25		2.5			
		0.25		-3		
0	0		-0.5		6	
		-0.25		3		0
1	-0.25		2.5		6	
		2.25		9		0
2	2		11.5		6	
		13.75		15		
3	15.75		20.5			
		40.25				
4	56					

Taking

$$x_0 = 0, y_0 = 0, \Delta y_0 = -0.25, \Delta^2 y_0 = 2.5, \Delta^3 y_0 = 9, \Delta^4 y_0 = 6,$$

Newton's forward difference formula for the first derivative gives,

$$\frac{dy}{dx} = \frac{1}{h}\left[\begin{array}{c} \Delta y_0 - \dfrac{2p-1}{2}\Delta^2 y_0 + \dfrac{3p^2 - 6p + 2}{6}\Delta^3 y_0 \\ -\dfrac{4p^3 - 18p^2 + 22p - 6}{24}\Delta^4 y_0 - \dots \end{array}\right]$$

$$= \frac{1}{1} - 0.25 + \frac{2x-1}{2}(2.5) + \frac{1}{6}\left(3x^2 - 6x + 2\right)9 +$$

$$\frac{1}{24}\left(4x^3 - 18x^2 + 22x - 6\right)6$$

$$= \frac{1}{2}\left[-0.25 + 2.5x - 1.25 + 4.5x^2 + 5.5x - 1.5\right] = x^3 - x$$

For y to be maximum or minimum, dy/dx=0, i.e., $x^3 - x = 0$

i.e., x=0, 1, -1

$$\frac{d^2 y}{dx^2} = 3x^2 - 1 = -ve \text{ for } x = 0$$

Now,
$$= +ve \text{ for } x = 1$$
$$= +ve \text{ for } x = -1,$$

Since $\qquad y = y_0 + x\Delta y_0 + \dfrac{x(x-1)}{2}\Delta^2 y_0 + \dots y(0) = 0$

Thus y is maximum for x=0, and minimum, value =y=0

Also y is minimum for x=1, and minimum value =y=-0.25

Exercises

1. Find $y'(0)$ and $y(0)$ from the following table

X	0	1	2	3	4	5
Y	4	8	15	7	6	2

2. Find the first, second and third derivative of y at x=1.5 if

X	1.5	2.0	2.5	3.0	3.5	4.0
Y	3.375	7.000	13.625	24.000	38.875	59.000

3. Find the first and second derivatives of the function tabulated below, at the point x=1.1

X	1.0	1.2	1.4	1.6	1.8	2.0
Y	0	0.128	0.544	1.296	2.432	4.0000

4. Given the following table of values of x and y

X	1.00	1.05	1.10	1.15	1.20	1.25	1.30
Y	1.00	1.02	1.04	1.07	1.09	1.11	1.14
	0	5	9	2	5	8	0

Find $\dfrac{dy}{dx}, \dfrac{d^2y}{dx^2}$ at $x = 1.05, x = 1.25, x = 1.15$

5. For the following values of x and y, find the first derivative at x=4.

X	1	2	4	8	10
Y	0	1	5	21	27

6. Find the derivative of y at x=0.4 from the following data

X	0.1	0.2	0.3	0.4
Y	1.10517	1.22140	1.34986	1.49182

7. From the following table, find the values of dy/dx and second derivative of y at x=2.03.

X	1.00	1.98	2.00	2.02	2.04
Y	0.7825	0.7739	0.7651	0.7563	0.7473

8. Given

$$\sin 0^0 = 0.000, \sin 10^0 = 0.1736, \sin 20^0 = 0.3420, \sin 30^0 = 0.5000,$$
$$\sin 40^0 = 0.6428$$

    i.   Find the value of sin23.

    ii.   Find the numerical value of cosx at x=10.

    iii.   Find the numerical value of second derivative at x=20 for y=sinx.

9. The population of a certain town is given below, find the rate of growth of the population in 1961 from the following table

Year	1931	1941	1951	1961	1971
Population	40.62	60.50	71.95	103.56	132.68

Estimate the population in the years 1976 and 2003. Also find the rate of growth of population in 1991.

10. The following data gives corresponding values of pressure and specific volume of a superheated stream.

V	2	4	6	8	10
P	100	42.7	25.3	16	7.13

Find the rate of change of i. pressure with respect to volume when v=2.

iii.   volume with respect to pressure when p=105.

11. The table below reveals the velocity v of a body during the specified at time t. find its acceleration at t=2.

T	1.0	1.1	1.2	1.3	1.4
C	43.1	47.7	52.1	56.4	60.8

12. The following table gives the velocity v of a particle at time t. Find its acceleration at t=2.

T	0	2	4	6	8	10	12
V	4	6	16	34	60	94	131

13. A rod is rotating in a plane. The following table gives the angle θ through which the rod has turned for various values of the time t second.

14. Find dy/dx at x=1 from the following table by constructing a central difference table

X	0.7	0.8	0.9	1.0	1.1	1.2	1.3
Y	0.64	0.71	0.78	0.84	0.89	0.93	0.96
	4218	7356	3327	1471	1207	2039	3558

15. Find the value of $f'(x)$ at x=0.04 from the following table using Bessels formula,

X	0.01	0.02	0.03	0.04	0.05	0.06
Y	0.102	0.104	0.107	0.109	0.112	0.114
	3	7	1	6	2	8

16. Find the value of $f'(8)$ from the table given below

X	6	7	9	42
Y	1.556	1.690	1.908	2.158

17. Give the following pairs of values of x and y

X	1	2	4	8	10
Y	0	1	5	21	27

Determine numerically dy/dx at x=4.

18. Find the maximum and minimum value of y from the following table

X	0	1	2	3	4	5
Y	0	0.25	0	2.25	16	56.25

19. Using the following data, find x for which y is minimum and find its value of y.

X	0	0.65	0.70	0.75
Y	0.6221	0.6155	0.6138	0.6170

20. Find the value of x for which y is maximum, using the table

X	9	10	11	12	13	14
Y	1330	1340	1320	1250	1120	930

Also find the maximum value of y.

4.3. Numerical integration

The process of evaluating a definite integral from a set of tabulated values of the integrand y is called *numerical integration*. This process when applied to a function of a single variable, is known as *quadrature*. The problem of numerical integration, like that of numerical differentiation is solved by representing $f(x)$ by an interpolation formula and then integrating it between the given limits. In this way, we can derive quadrature formulae for approximate integration of a function defined by a set of numerical values only.

Let $y = f(x)$ be a function on [a, b]. If the function $f(x)$ is defined explicitly and the integral of $f(x)$ can be calculated by the usual methods then the definite integral $\int_a^b f(x)\,dx$ can be found easily.

211

If the function is given in tabular form and the integral of $f(x)$ is different to find then numerical integration is needed. Numerical integration is used to obtain approximate answers for definite integrals that cannot be solved analytically. It is a process of finding the numerical value of a definite integral.

$$I = \int_a^b f(x)\,dx \text{ when the function } y = f(x) \text{ is not known}$$

explicitly.

Let $y = f(x)$ be a tabulated function i.e., let $y_0, y_1, y_2, \ldots y_n$ be the values of $f(x)$ for $x = x_0, x_1, x_2, \ldots x_n$. Then we replace $f(x)$ with an interpolation polynomial and then integrate it with in the desired limits. Here we integrate an approximate interpolation formula instead of $f(x)$. When this method is applied on a function of single variable, the process is called quadrature.

$$\text{Let } I = \int_a^b f(x)\,dx$$

For $f(x)$, we fit an approximate polynomial $\phi(x)$ of suitable degree with in [a, b]   i.e.,

$$\int_a^b f(x)\,dx \cong \int_a^b \phi(x)\,dx$$

The difference $\int_a^b f(x)\,dx - \int_a^b \phi(x)\,dx$ is called the error of approximation.

4.4. Newton-cotes Quadrature Formula

Let $y = f(x)$ be a function in [a,b]. At $x = x_0, x_1, x_2, \ldots x_n$ let the corresponding values of y be

$y_0, y_1, y_2, \ldots y_n$. Let us divide the interval (a,b) into n sub-intervals of width h so that $x_0 = a, x_1 = x_0 + h, x_2 = x_0 + 2h = b$. Then,

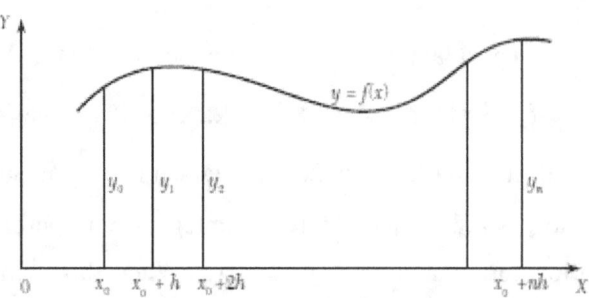

Fig.4.1 Newton-cotes Quadrature Formula

$$I = \int_{x_0}^{x_0 + nh} f(x)\, dx$$

$$== h\int_0^n f(x_0 + rh)\, dr,\ x = x_0 + rh,\ dx = hdr$$

$$= h\int_0^n \left[\begin{array}{l} y_0 + r\Delta y_0 + \dfrac{r(r-1)}{2}\Delta^2 y_0 + \dfrac{r(r-1)(r-2)}{6}\Delta^3 y_0 \\[2mm] + \dfrac{r(r-1)(r-2)(r-3)}{24}\Delta^4 y_0 + \ldots \end{array} \right] dr$$

By Newton's forward interpolation formula. Integrating term by term, we obtain

$$\int_{x_0}^{x_0+nh} f(x)\,dx = nh \left[\begin{array}{l} y_0 + \dfrac{n}{2}\Delta y_0 + \dfrac{n(2n-3)}{12}\Delta^2 y_0 \\[2mm] + \dfrac{n(n-2)^2}{24}\Delta^3 y_0 \\[2mm] + \left(\dfrac{n^4}{5} - \dfrac{3n^3}{2} + \dfrac{11n^2}{3} - 3n \right)\dfrac{\Delta^4 y_0}{24} + \\[2mm] \left(\dfrac{n^5}{6} - 2n^4 + \dfrac{34n^3}{4} - \dfrac{50n^2}{3} + 12n \right)\dfrac{\Delta^5 y_0}{120} \\[2mm] +.... \end{array} \right]$$

This is known as *Newton-Cotes quadrature formula.*

From this general formula, we deduce the following important quadrature rules by taking $n = 1, 2, 3$, Here x values are equally spaced with common difference h.

Newton's forward difference formula is

$$y = y_0 + p\Delta y_0 + \frac{p(p-1)}{\angle 2}\Delta^2 y_0 + \frac{p(p-1)(p-2)}{\angle 3}\Delta^3 y_0 + ...$$

where $p = \dfrac{x - x_0}{h} \Rightarrow x = x_0 + ph$

Then $\qquad \displaystyle\int_a^b f(x)\,dx = \int_{x_0}^{x_n} f(x)\,dx$

214

$$\int_{x_0}^{x_n} y\,dx = \int_{x_0}^{x_n} \left[y_0 + p\Delta y_0 + \frac{p^2 - p}{\angle 2}\Delta^2 y_0 + \left[\frac{p^3 - 3p^2 + 2p}{6}\right]\Delta^3 y_0 + \ldots \right] dx$$

We have $p = \dfrac{x - x_0}{h} \Rightarrow x = x_0 + ph$

$$\Rightarrow \frac{dp}{dx} = \frac{1}{h}$$

$$\Rightarrow h.dp = dx$$

If $x = x_0 \Rightarrow p = 0$

If $x = x_n \Rightarrow p = \dfrac{x_n - x_0}{h} = n$

As x varies from x_0 to x_n P varies from 0 to n.

$$\Rightarrow \int_{x_0}^{x_n} y\,dx = \int_0^n \left(\begin{array}{l} y_0 + p\Delta y_0 + \dfrac{p(p-1)}{\angle 2}\Delta^2 y_0 \\[2mm] + \dfrac{p^3 - 3p^2 + 2p}{\angle 3}\Delta^3 y_0 + \ldots \end{array} \right)$$

$$= h \left[\begin{array}{l} py_0 + \dfrac{p^2}{2}\Delta y_0 + \left(\dfrac{p^3}{3} - \dfrac{p^2}{2}\right)\dfrac{\Delta^2 y_0}{2} \\[3mm] + \left(\dfrac{p^4}{4} - p^3 + p^2\right)\dfrac{\Delta^3 y_0}{6} + \ldots \end{array} \right]_0^n$$

$$= h \left[ny_0 + \frac{n^2}{2}\Delta y_0 + \left(\frac{n^3}{3} - \frac{n^2}{2} \right)\frac{\Delta^2 y_0}{2} \right.$$
$$\left. + \left(\frac{n^4}{4} - n^3 + n^2 \right)\frac{\Delta^3 y_0}{6} + \ldots \right]$$

$$\Rightarrow \int_{x_0}^{x_n} y\,dx = h \left[ny_0 + \frac{n^2}{2}\Delta y_0 + \left(\frac{n^3}{3} - \frac{n^2}{2} \right)\frac{\Delta^2 y_0}{2} \right.$$
$$\left. + \left(\frac{n^4}{4} - n^3 + n^2 \right)\frac{\Delta^3 y_0}{6} + \ldots \right]$$

$$\Rightarrow \int_{x_0}^{x_n} y\,dx = nh \left[y_0 + \frac{n}{2}\Delta y_0 + \left(\frac{2n^2 - 3n}{12} \right)\Delta^2 y_0 \right.$$
$$\left. + \left(\frac{n^3 - 4n^2 + 4n}{24} \right)\Delta^3 y_0 + \ldots \right]$$

This is called the general quadrature formula for equidistant ordinates, from which we can generate any numerical integration formula.

4.4.1. TRAPEZOIDAL RULE

Put n=1, then all differences higher than the first will become zero and we obtain from Newton's quadrature formula.

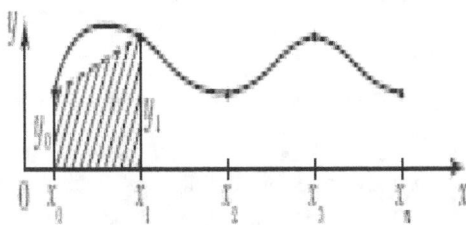

Fig.4.2 Trapezoidal Rule

216

$$\int_{x_0}^{x_1} y dx = h\left[y_0 + \frac{1}{2}\Delta y_0\right]$$

$$= h\left[y_0 + \frac{1}{2}(y_1 - y_0)\right]$$

$$= \frac{h}{2}[y_0 + y_1]$$

In the next interval $\left[x_1, x_2\right]$ we obtain

$$\int_{x_1}^{x_2} y dx = \frac{h}{2}[y_1 + y_2]$$

$$\int_{x_2}^{x_1} y dx = \frac{h}{2}[y_2 + y_3]$$

and so on.

In the interval $\left[x_{n-1}, x_n\right]$, we get $\int_{x_{n-1}}^{x_n} y dx = \frac{h}{2}[y_{n-1} + y_n]$

Adding all these equations

$$\int_{x_0}^{x_n} y dx = \frac{h}{2}[y_0 + y_1 + y_2 + \ldots\ldots y_{n-1}, y_n]$$

$$= \frac{h}{2}\left[y_0 + y_n + 2(y_1 + y_2 + y_3 + \ldots. + y_{n-1})\right]$$

This is known as trapezoidal rule.

Geometrical significance

Let $y = f(x)$ be a curve. Let $y_0, y_1, y_2, \ldots y_n$ be the values of y corresponding to $x = x_0, x_1, x_2, \ldots x_n$. Let the curve $y = f(x)$ pass through $(x_0, y_0), (x_1, y_1), (x_2, y_2) \ldots (x_n, y_n)$. The wave $y = f(x)$ can be replaced by n straight lines joining the points $(x_0, y_0), (x_1, y_1), (x_2, y_2) \ldots (x_n, y_n)$. Then the area bounded by the curve $y = f(x)$, the ordinates $x = x_0, x = x_n$ and x-axis is equivalent to the sum of areas of n-trapeziums formed.

$$\int_{x_0}^{x_n} y \, dx = \frac{h}{2} \left[y_0 + y_n + 2(y_1 + y_2 + y_3 + \ldots + y_{n-1}) \right]$$

4.4.2. Simpson's one third rule

Put n=2, in Newton's general quadrature then all differences higher than the first will become zero and we obtain

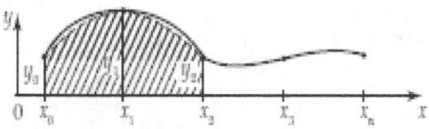

Fig.4.3 Simpson's one third rule

$$\int_{x_0}^{x_1} ydx = 2h\left[y_0 + \frac{2}{2}\Delta y_0 + \frac{2.2^2 - 3.2}{12}\Delta^2 y_0\right]$$

$$= 2h\left[y_0 + \Delta y_0 + \frac{1}{6}\Delta^2 y_0\right]$$

$$= 2h\left[y_0 + y_1 - y_0 + \frac{1}{6}(y_2 - 2y_1 + y_0)\right]$$

$$= 2h\left[\frac{6y_1 + y_2 - 2y_1 + y_0}{6}\right]$$

$$= \frac{h}{3}[4y_1 + y_2 + y_0] = \frac{h}{3}[y_0 + 4y_1 + y_2]$$

$$\int_{x_2}^{x_4} ydx = \frac{h}{3}[y_2 + 4y_3 + y_4]$$

Similarly,

$$\int_{x_4}^{x_6} ydx = \frac{h}{3}[y_4 + 4y_5 + y_6]$$

$$\int_{x_2}^{x_4} ydx = \frac{h}{3}[y_2 + 4y_3 + y_4]$$

$$\int_{x_4}^{x_6} ydx = \frac{h}{3}[y_4 + 4y_5 + y_6]$$

$$\int_{x_{n-2}}^{x_n} ydx = \frac{h}{3}[y_{n-2} + 4y_{n-1} + y_n]$$

Adding all these equations

$$\int_{x_0}^{x_n} y\,dx = \frac{h}{3}\left[\begin{array}{l} y_0 + 4y_1 + y_2 + y_2 + 4y_3 + y_4 + y_4 \\ +\ldots\ldots y_{n-2} + 4y_{n-1} + y_n \end{array}\right]$$

$$= \frac{h}{3}\left[\begin{array}{l} y_0 + y_n + 4\left(y_1 + y_2 + y_3 + \ldots + y_{n-1}\right) \\ +2\left(y_2 + y_4 + \ldots + y_{n-2}\right) \end{array}\right]$$

$$\Rightarrow \int_{x_0}^{x_n} y\,dx = \frac{h}{3}\left[\begin{array}{l} y_0 + y_n + 4\left(\text{sum of odd ordinates}\right) \\ +2\left(\text{sum of even ordinates}\right) \end{array}\right]$$

This is known as Simpson's 1/3rd rule.

Note. It can be applied only when the given interval [a, b] is subdivided into even number of subintervals each of width h and within any two consecutive subintervals the interpolating polynomial $\phi(x)$ is of degree.

4.4.3. Simpson's 3/8th rule

Here we assume that within any three consecutives subintervals of width h, the interpolating polynomial $\phi(x)$ is of degree 3. Hence put n=3 in 1 and neglecting the fourth and higher order differences, then

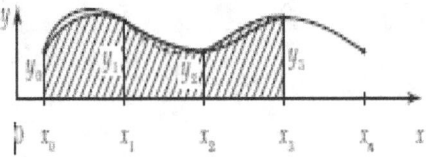

Fig.4.4 Simpson's 3/8th rule

$$\int_{x_0}^{x_n} y dx = 3h \left[\begin{array}{l} y_0 + \dfrac{3}{2}\Delta y_0 + \dfrac{2.3^2 - 3.3}{12}\Delta^2 y_0 \\ + \dfrac{3^3 - 4.3^2 + 4.3}{24}\Delta^3 y_0 \end{array} \right]$$

$$= 3h \left[y_0 + \dfrac{3}{2}\Delta y_0 + \dfrac{3}{4}\Delta^2 y_0 + \dfrac{3}{24}\Delta^3 y_0 \right]$$

$$= 3h \left[\begin{array}{l} y_0 + \dfrac{3}{2}(y_1 - y_0) + \dfrac{3}{4}(y_2 - 2y_1 + y_0) \\ + \dfrac{3}{24}(y_3 - 3y_2 + 3y_1 - y_0) \end{array} \right]$$

$$= \dfrac{3h}{8}\left[y_0 + 3y_1 + 3y_2 + y_3 \right]$$

Similarly,

$$\int_{x_3}^{x_6} y dx = \dfrac{3h}{8}\left[y_3 + 3y_4 + 3y_5 + y_6 \right]$$

$$\int_{x_6}^{x_9} y dx = \dfrac{3h}{8}\left[y_6 + 3y_7 + 3y_8 + y_9 \right]......$$

Finally, $\int_{x_{n-3}}^{x_n} y dx = \dfrac{3h}{8}\left[y_{n-3} + 3y_{n-2} + 3y_{n-1} + y_n \right]$

Adding all these equations,

$$\int_{x_0}^{x_n} y dx = \dfrac{3h}{8}\left[\begin{array}{l} y_0 + 3y_1 + 3y_2 + y_3 + y_3 + 3y_4 + 3y_5 + y_6 + ... \\ + y_{n-3} + 3y_{n-2} + 3y_{n-1} + y_n \end{array} \right]$$

$$= \dfrac{3h}{8}\left[\begin{array}{l} (y_0 + y_n) + 3(y_1 + y_2 + y_4 + y_5 + .. + y_{n-1}) \\ + 2(y_3 + y_6 + ... + y_{n-3}) \end{array} \right]$$

This is called Simpson's 3/8th rule.

It can be applied when the range [a, b] is divided into a number of subintervals, which is a multiple of 3.

Example.4.11 Evaluate $\int_{1}^{3}\dfrac{1}{x}dx$ by Simpson's rule with 4 strips and 8 strips respectively. Determine the error by direct integration.

Solution Here $x_0 = 1, x_n = 3, y = \dfrac{1}{x}$

With 4 strips

$$n = 4, h = \frac{x_n - x_0}{n} = \frac{3-1}{4} = 0.5$$

x	$x_0 = 1$	$x_1 = 1.5$	$x_2 = 2$	$x_3 = 2.5$	$x_4 = 3$
$y = \dfrac{1}{x}$	$y_0 = 1$	$y_1 = 0.6666$	$y_2 = 0.5$	$y_3 = 0.4$	$y_4 = 0.333$

Simpson's 1/3rd rule,

$$\int_{x_0}^{x_n} ydx = \frac{h}{3}\left[y_0 + y_n + 4(y_1 + y_3 + ...) + 2(y_2 + y_4 + ...) \right]$$

$$= \frac{h}{3}\left[y_0 + y_4 + 4(y_1 + y_3) + 2y_2 \right]$$

$$= \frac{0.5}{3}\left[1 + 0.333 + 4(0.666 + 0.4) + 2(0.5) \right]$$

$$= \frac{0.5}{3}\left[1.333 + 4.264 + 1 \right]$$

$$= 1.0995$$

222

With 8 strip

$$n = 8, h = \frac{x_n - x_0}{n} = \frac{3-1}{8} = \frac{1}{4} = 0.25$$

x	1	1.25	1.5	1.75	2	2.25	2.5	2.75	3
y	1	0.8	0.666	0.571	0.5	0.444	0.4	0.3636	0.333

Simpson's 1/3rd rule

$$\int_{x_0}^{x_n} ydx = \frac{h}{3}\left[y_0 + y_n + 4(y_1 + y_3 + ...) + 2(y_2 + y_4 + ...)\right]$$

$$= \frac{h}{3}\left[y_0 + y_8 + 4(y_1 + y_3 + y_5 + y_7) + 2(y_2 + y_4 + y_6)\right]$$

$$= \frac{0.25}{3}\left[\begin{array}{l}1 + 0.333 + 4(0.8 + 0.571 + 0.444 + 0.3636) \\ +2(0.666 + 0.5 + 0.4)\end{array}\right]$$

$$= \frac{0.25}{3}[1.333 + 8.7144 + 3.132]$$

$$= 1.09828$$

By direct integration,

$$\int_1^3 \frac{1}{x} dx = (\log x)_1^3 = \log 3 - \log 1 = 1.0986.$$

Absolute Error=Exact Value - Obtained value.

Error due to 4 strips=1.0986-1.0995=-0.0009.

Error due to 8 strips=1.0986-1.09828=0.00032.

Example.4.12 Calculate the value $\int_0^1 \dfrac{x}{x+1} dx$ correct to 3 significant figures taking 6 intervals by trapezoidal rule.

Solution Let $x_0 = 0, x_n = 1, n = 6, h = \dfrac{x_n - x_0}{n} = \dfrac{1-0}{6} = \dfrac{1}{6}$

X	0	1/6	1/3	1/2	2/3	5/6	1
y	0	0.1428	0.25	0.333	0.4	0.4545	0.5

By trapezoidal rule,

$$\int_{x_0}^{x_n} y\,dx = \frac{h}{2}\left[y_0 + y_n + 2\left(y_1 + y_2 + y_3 + ... + y_{n-1} \right)\right]$$

$$\int_0^1 \frac{x}{1+x} dx = \frac{h}{2}\left[y_0 + y_6 + 2\left(y_1 + y_2 + y_3 + y_4 + y_5 \right)\right]$$

$$= \frac{1}{12}\left[0 + 0.5 + 2\left(0.1428 + 0.25 + 0.333 + 0.4 + 0.4545 \right)\right]$$

$$= 0.30505$$

Example.4.13 Find the value of $\int_0^1 \dfrac{1}{1+x^2} dx$ taking 5 subintervals by trapezoidal rule. Correct to 4 significant figures also compare it with its exact value.

Solution Let $x_0 = 0, x_n = 1, y = \dfrac{1}{1+x^2}, n = 5$

$$h = \frac{x_n - x_0}{n} = \frac{1-0}{5} = 0.2$$

X	0	0.2	0.4	0.6	0.8	1
Y	1	0.9615	0.8620	0.7352	0.6097	0.5

Trapezoidal rule,

$$\int_{x_0}^{x_n} y\,dx = \frac{h}{2}\Big[y_0 + y_n + 2\big(y_1 + y_2 + y_3 + \dots + y_{n-1}\big)\Big]$$

$$\int_0^1 \frac{1}{1+x^2}\,dx = \frac{h}{2}\Big[y_0 + y_5 + 2\big(y_1 + y_2 + y_3 + y_4\big)\Big]$$

$$= \frac{0.2}{2}\Big[1 + 0.5 + 2\big(0.9615 + 0.8620 + 0.7352 + 0.6097\big)\Big]$$

$$= 0.78368$$

By direct integration,

$$\int_0^1 \frac{1}{1+x^2}\,dx = \big(Tan^{-1}x\big)_0^1 = \frac{\pi}{4} - 0 = 0.78539$$

Error=Exact value-obtained value

=0.78539-0.78368=0.00171

Example.4.14 Find the approximate value of

$$\int_0^{\pi/2} \sqrt{\cos x}\,dx \text{ by dividing the interval into 4 parts.}$$

Solution Given that

$$x_0 = 0, x_n = \frac{\pi}{2}, y = \sqrt{\cos x}, n = 4,$$

$$h = \frac{x_n - x_0}{n} = \frac{\pi}{8}$$

X	y
0	1
$\dfrac{\pi}{8}$	0.9611
$\dfrac{\pi}{4}$	0.8408
$\dfrac{3\pi}{8}$	0.6186
$\dfrac{\pi}{2}$	0

By Trapezoidal rule,

$$\int_0^{\pi/2} \sqrt{\cos x}\,dx = \frac{h}{3}\Big[y_0 + y_4 + 4(y_1 + y_3) + 2y_2\Big]$$

$$= \frac{\pi}{24}\Big[1 + 0 + 4(0.9611 + 0.6186) + 2(0.8408)\Big]$$

$$= \frac{\pi}{24}\Big[1 + 6.3188 + 1.6816\Big]$$

$$= 1.17814$$

Example. 4.15 Determine the maximum error in evaluating $\int_0^{\pi/2} \cos x\,dx$ by trapezoidal rule and Simpson's rules using 4 subintervals.

Solution Let $x_0 = 0, x_n = \dfrac{\pi}{2}, y = \cos x, n = 4, h = \dfrac{x_n - x_0}{n} = \dfrac{\pi}{8}$

x	y
0	1
$\dfrac{\pi}{8}$	0.9238
$\dfrac{\pi}{4}$	0.7071
$\dfrac{3\pi}{8}$	0.3826
$\dfrac{\pi}{2}$	0

Trapezoidal rule,

$$\int_{x_0}^{x_n} y dx = \frac{h}{2}\left[y_0 + y_n + 2\left(y_1 + y_2 + y_3 + \ldots + y_{n-1} \right) \right]$$

$$\int_{0}^{\pi/2} \cos x dx = \frac{h}{2}\left[y_0 + y_4 + 2\left(y_1 + y_2 + y_3 \right) \right]$$

$$= \frac{\pi}{16}\left[1 + 0 + 2\left(0.9238 + 0.7071 + 0.3826 \right) \right]$$

$$= 0.98704$$

227

By Simpson's $1/3^{rd}$ rule

$$\int_{x_0}^{x_n} ydx = \frac{h}{3}\left[y_0 + y_n + 4\left(y_1 + y_3 + ...\right) + 2\left(y_2 + y_4 + ...\right)\right]$$

$$= \frac{h}{3}\left[y_0 + y_4 + 4\left(y_1 + y_3\right) + 2\left(y_2\right)\right]$$

$$= \frac{\pi}{24}\left[1 + 0 + 4\left(0.9238 + 0.3826\right) + 2\left(0.7071\right)\right]$$

$$= \frac{\pi}{24}\left[1 + 5.2256 + 1.4142\right]$$

$$= 1.000047$$

By direct integration,

$$\int_0^{\pi/2} \cos xdx = \left(\sin x\right)_0^{\pi/2} = \sin\frac{\pi}{2} - \sin 0 = 1$$

Error in Trapezoidal rule =1-0.98704=0.01296.

Error in Simpson's rule =1-1.000047= -0.000047.

Maximum Error=0.01296

Example.4.16. Calculate $\int_0^{\pi/2} e^{\sin x}dx$ correct to 4 decimal places

Solution Let $x_0 = 0, x_n = \frac{\pi}{2}, y = e^{\sin x}, n = 4, h = \frac{x_n - x_0}{n} = \frac{\pi}{8}$

X	Y
0	1
$\dfrac{\pi}{8}$	1.4662
$\dfrac{\pi}{4}$	2.0281
$\dfrac{3\pi}{8}$	2.5190
$\dfrac{\pi}{2}$	2.7182

Simpson's 1/3rd rule

$$\int_{x_0}^{x_n} y\,dx = \frac{h}{3}\Big[y_0 + y_n + 4(y_1 + y_3 + ...) + 2(y_2 + y_4 + ...)\Big]$$

$$\int_{0}^{\pi/2} e^{\sin x}\,dx = \frac{h}{3}\Big[y_0 + y_4 + 4(y_1 + y_3) + 2(y_2)\Big]$$

$$= \frac{\pi}{24}\Big[1 + 2.7182 + 4(1.4662 + 2.5190) + 2(2.0281)\Big]$$
$$= 3.1043$$

Example.4.17 Evaluate $\displaystyle\int_{0}^{6} \frac{1}{1+x^2}\,dx$ using Simpson's 3/8th rule and Simpson's 1/3rd rule.

Solution Given $x_0 = 0, x_n = 6, y = \dfrac{1}{1+x^2}, n = 6$

$$h = \frac{x_n - x_0}{n} = \frac{6-0}{6} = 1$$

x	$y = \dfrac{1}{1+x^2}$
0	1
1	0.5
2	0.2
3	0.1
4	0.0588
5	0.0384
6	0.0270

Simpson's 1/3rd rule

$$\int_{x_0}^{x_n} y\,dx = \frac{h}{3}\left[y_0 + y_n + 4(y_1 + y_3 + ...) + 2(y_2 + y_4 + ...)\right]$$

$$\int_0^6 \frac{1}{1+x^2}\,dx = \frac{h}{3}\left[y_0 + y_6 + 4(y_1 + y_3 + y_5) + 2(y_2 + y_4)\right]$$

$$= \frac{1}{3}\left[1 + 0.0270 + 4(0.5 + 0.1 + 0.0384) + 2(0.2 + 0.0588)\right]$$

$$= \frac{1}{3}\left[1.0270 + 2.5536 + 0.5176\right]$$

$$= 1.3660$$

Simpson's 3/8th rule

$$\int_{x_0}^{x_n} ydx = \frac{3h}{8}\left[\begin{array}{c}(y_0 + y_n) + 3(y_1 + y_2 + y_4 + y_5 + .. + y_{n-1}) \\ +2(y_3 + y_6 + ... + y_{n-3})\end{array}\right]$$

$$\int_0^6 \frac{1}{1+x^2}dx = \frac{3h}{8}\left[y_0 + y_6 + 3(y_1 + y_2 + y_4 + y_5) + 2y_3\right]$$

$$= \frac{3}{8}\left[1 + 0.0270 + 3(0.5 + 0.2 + 0.0588 + 0.0384) + 2(0.1)\right]$$

$$= \frac{3}{8}\left[1.0270 + 2.3916 + 0.2\right]$$

$$= 1.3569$$

Example.4.18 The velocity of a train which starts from rest is given by the following table

T min	2	4	6	8	10	12	14	16	18	20
V km/hr	16	28.8	40	46.4	51.2	32	17.6	8	3.2	0

Estimate total distance run in 20min.

Solution We know that,

$$V = \frac{ds}{dt} \Rightarrow ds = vdt; s = \int_0^{20} vdt$$

Since the train starts from rest $\Rightarrow v = 0$ at t=0,

$$t_0 = 0, t_n = 20, h = \frac{20-0}{10} = \frac{2}{1} = \frac{2}{60} hrs = \frac{1}{30} hrs$$

Simpson's $1/3^{rd}$ rule

$$\int_0^{20} v\,dt = \frac{h}{3}\left[\begin{array}{l} y_0 + y_{10} + 4\left(y_1 + y_3 + y_5 + y_7 + y_9\right) \\ +2\left(y_2 + y_4 + y_6 + y_8\right) \end{array}\right]$$

$$= \frac{1}{90}\left(\begin{array}{l} 0+0+4\left(16+40+51.2+17.6+3.2\right) \\ +2\left(28.8+46.4+32+8\right) \end{array}\right)$$

$$= 8.25 km$$

The distance run by train in 20min=8.25km.

Example.4.19 Find the value of $\int_1^5 \log_{10}^x dx$ taking 8 subintervals correct to 4 decimal places by trapezoidal rule.

Solution   Given

$$x_0 = 1, x_n = 5, n = 8, h = \frac{x_n - x_0}{n} = \frac{5-1}{8} = 0.5, y = \log_{10}^x$$

X	y
1	0
1.5	0.17609
2	0.30103
2.5	0.39794
3	0.47712
3.5	0.54407
4	0.60206
4.5	0.65321

232

5	0.69897

By Trapezoidal rule,

$$\int_{x_0}^{x_n} y\,dx = \frac{h}{2}\left[y_0 + y_n + 2\left(y_1 + y_2 + y_3 + \ldots + y_{n-1}\right)\right]$$

$$\int_{1}^{5} \log_{10}^{x} dx = \frac{h}{2}\left[y_0 + y_8 + 2\left(y_1 + y_2 + y_3 + y_4 + y_5 + y_6 + y_7\right)\right]$$

$$= \frac{0.5}{2}\left[0 + 0.69897 + 2\begin{pmatrix} 0.17609 + 0.30103 + 0.39794 \\ +0.47712 + 0.54407 + 0.60206 \\ +0.65321 \end{pmatrix}\right]$$

$$= 1.7505025$$

4.4.4. Applications of Simpson's rule. If the various ordinates in sections represent equispaced cross-sectional areas, thenSimpson's rule gives the volume of the solid. As such, Simpson's rule is very useful to civil engineers for calculating the amount of earth that must be moved to fill a depression or make a dam.

Similarly, if the ordinates denote velocities at equal intervals of time, Simpson's rule gives the distance traveled. The following Examples illustrate these applications.

Exercise

1. Use trapezoidal rule to evaluate $\int_0^1 x^2 dx$ considering five sub intervals.

2. Evaluate $\int_0^1 \frac{1}{1+x} dx$ applying Trapezoidal rule, Simpson's 1/3rd rule, Simpsons 3/8th rule.

3. Find an approximate value of $\log_e^5$ by calculating to 4 decimal places, by Simpson's 1/3rd rule, $\int_0^5 \frac{dx}{4x+5}$ dividing the range into ten equal parts.

4. Find $\int_0^6 \frac{e^x}{1+x} dx$ using Simpson's 1/3rd rule.

5. Evaluate $\int_0^2 e^{-x^2} dx$ using Simpson's rule.

6. Evaluate using Simpson's 1/3rd rule

    i. $\int_0^\pi \sin x dx$, taking eleven ordinates.

    ii. $\int_0^{\frac{\pi}{2}} \sqrt{\cos\theta} d\theta$ taking nine ordinates

7. Evaluate by Simpson's 3/8th rule

    i. $\int_0^9 \frac{1}{1+x^2} dx$

    ii. $\int_0^{\frac{\pi}{2}} \sin x dx$

iii. $\displaystyle\int_{0}^{\frac{\pi}{2}} e^{\sin x}\,dx$

iv. $\displaystyle\int_{0}^{\pi} \sqrt{\left(1+3\cos^2\theta\right)}\,d\theta$ using 6 ordinates.

8. Given that

X	4.0	4.2	4.4	4.6	4.8	5.0	5.2
Lo	1.38	1.43	1.48	1.52	1.56	1.60	1.64
gx	63	51	16	61	86	94	87

Evaluate $\displaystyle\int_{4}^{5.2} \log x\,dx$

i. Trapezoidal rule
ii. Simpson's $1/3^{\text{rd}}$ rule
iii. Simpson's $3/8^{\text{th}}$ rule

4.5. ROMBERG'S METHOD

we have derived approximate quadrature formulae with the help of finite differences method. Romberg's method provides a simple

modification to these quadrature formulae for finding their better approximations. As an illustration, let us improve upon the value of the integral

$$I = \int_{a}^{b} f(x)\,dx$$

By the trapezoidal rule. If I_1, I_2 are the values of I with subintervals of width h_1, h_2 and E_1, E_2 their corresponding errors, respectively, then

235

$$E_1 = \frac{-(b-a)h_1^2}{12} y''(X), E_2 = \frac{-(b-a)h_2^2}{12} y''(\overline{X})$$

Since $y''(\overline{X})$ is also the largest value of $y''(x)$,

we can reasonably assume that $y''(x)$ and $y''(\overline{X})$ are very nearly equal.

$$\therefore \frac{E_1}{E_2} = \frac{h_1^2}{h_2^2} \quad \text{or}$$

$$\frac{E_1}{E_2 - E_1} = \frac{h_1^2}{h_2^2 - h_1^2}$$
$$\because I = I_1 + E_1 = I_2 + E_2$$
$$\therefore E_2 - E_1 = I_1 - I_2$$

From above equations

$$E_1 = \frac{h_1^2}{h_2^2 - h_1^2}(I_1 - I_2)$$

$$I = I_1 + E_1 = I_1 + \frac{h_1^2}{h_2^2 - h_1^2}(I_1 - I_2)$$

$$i.e., I = \frac{I_1 h_2^2 - I_2 h_1^2}{h_2^2 - h_1^2}$$

Which is a better approximation of I.

To evaluate I systematically, we take $h_1 = h$ and $h_2 = \frac{h}{2}$

So that gives

$$I = \frac{I_1 (h/2)^2 - I_2 h_2^2}{(h/2)^2 - h^2} = \frac{4I_2 - I_1}{3}$$

$$i.e., I(h, h/2) = \frac{1}{3}\left(4I(h/2) - I(h)\right)$$

Now we use the trapezoidal rule several times successively having h and apply to each pair of values as per the following scheme

$I(h)$			
	$I(h, h/2)$		
$I(h/2)$		$I(h, h/2, h/4)$	
	$I(h/2, h/4)$		$I(h, h/2, h/4, h/8)$
$I(h/4)$		$I(h/2, h/4, h/8)$	
	$I(h/4, h/8)$		
$I(h/8)$			

The computation is continued until successive values are close to each other.

This method is called *Richardson's deferred approach to the limit* and its systematic refinement is called *Romberg's method.*

Example. 4.20 Evaluate $\int_0^1 \dfrac{1}{1+x}\,dx$ correct to three decimal

places using Romberg's method. Hence find the value of $\log_e^2$.

Solution Taking h=0.5, 0.25, and 0.125 successively, let us
evaluate the given integral by the trapezoidal rule.

i.   When h=0.5, the values of $y = (1+x)^{-1}$ are

X	0	0.5	1
Y	1	0.6666	0.5

$$\therefore I = \frac{0.5}{2}\left(1+0.5+2*0.6666\right) = 0.7083$$

ii.   When h=0.25, the values of $y = (1+x)^{-1}$ are

X	0	0.25	0.5	0.75	1
Y	1	0.8	0.6666	0.5714	0.5

$$\therefore I = \frac{0.125}{2}\left[\left(1+0.5\right)+2\binom{0.8889+0.8+0.7272+0.6667}{+0.6513+0.5714+0.5333}\right]$$
$$= 0.6941$$

Using Romberg's formulae, we obtain

$$I(h.h/2) = \frac{1}{3}\left[4I(h/2) - I(h)\right]$$

$$= \frac{1}{3}\left[4*0.697 - 0.7083\right]$$

$$= 0.6932$$

$$I(h/2, h/4) = \frac{1}{3}\left[4I(h/4) - I(h/2)\right]$$

$$= \frac{1}{3}\left[4*0.6941 - 0.697\right]$$

$$= 0.6931$$

$$I(h, h/2, h/4) = \frac{1}{3}\left[4I(h/2) - I(h/4) - I(h, h/2)\right]$$

$$= 0.6931$$

Hence the value of the integral

$$\int_0^1 \frac{1}{1+x} dx = 0.693$$

$$\int_0^1 \frac{1}{1+x} dx = \left|\log(1+x)\right|_0^1 = \log 2$$

Hence $\log_e^2 = 0.693$

Example. 4.21 Use Romberg's method to compute

$$\int_0^1 \frac{1}{1+x^2} dx \text{ correct to 4 decimal places.}$$

Solution We take h=0.5,0.25, and 0.125 successively

and evaluate the given integral using the trapezoidal rule

i.   When h=0.5 the values of $y = (1+x)^{-1}$ are

X	0	0.5	1
Y	1	0.8	0.5

$$\therefore I = \frac{0.5}{2}[1+2*0.8*0.5] = 0.775$$

ii.   When h=0.25, the values of $y = (1+x)^{-1}$ are

X	0	0.25	0.5	0.75	1
Y	1	0.0412	0.8	0.64	0.5

$$\therefore I = \frac{0.25}{2}\left[1+2(0.9412+0.8+0.64)+0.5\right]$$

$$= 0.7828$$

iii.   When h=0.125, we find that I=0.7848

Thus we have

$$I(h) = 0.7750, I(h/2) = 0.7828, I(h/4) = 0.7848$$

Now using above, we obtain

$$I(h,h/2) = \frac{1}{3}\left[4I(h/2)-I(h)\right] = \frac{1}{3}(3.1312-0.775)$$

$$= 0.7854$$

$$I(h/2,h/4) = \frac{1}{3}\left[\left(4I(h/4)-I(h/2)\right)\right]$$

$$= \frac{1}{2}(3.142-0.7854) = 0.7855$$

Therefore, the table of these values is

0.7750		
	0.7854	
0.7828		0.7855
	0.7855	
0.7848		

Hence the value of the integral=0.7855.

Example.4.22   Evaluate the integral using Romberg's method, correct to three decimal places.

Solution Taking h=0.25, 0.125, 0.0625 successively, let us evaluate the given integral by using Simpson's $1/3^{rd}$ rule.

i.   When h=0.25 the values of $y = \dfrac{x}{sinx}$ are

X	0	0.25	0.5
Y	1	1.0105	1.0429
Y_0	Y_1	Y_2	Y_3

Therefore, By Simpson's rule,

$$I = \frac{h}{3}\left[(y_0 + y_2) + 4y_1\right] = \frac{0.25}{3}\left[(1+1.0429) + 1.0105\right]$$
$$= 0.5071$$

ii.   When h=0.125, the values of y are

X	0	0.125	0.25	0.375	0.5
Y	1	1.0026	1.0105	1.1003	1.0429
	Y_0	Y_1	Y_2	Y_3	Y_4

.

By Simpson's rule

$$I = \frac{h}{3}\left[(y_0 + y_4) + 4(y_1 + y_2) + 2y_2\right]$$

$$= \frac{0.125}{3}\left[\begin{array}{l}(1+1.0429) + 4(1.0026+1.1003)\\+2(1.0105)\end{array}\right]$$

$$= 0.5198$$

iii.   When h=0.0625, the values of y are

X	0	0.0625	0.125	0.1875	0.25	0.3125	0.1875	0.4375	0.5
Y	1	0.0006	1.0026	1.0059	1.0157	1.0165	1.1003	1.0326	1.0429
	y_0	y_1	y_2	y_3	y_4	y_5	y_6	y_7	y_8

By Simpson's rule

$$I = \frac{h}{3}\left[(y_0 + y_8) + 4(y_1 + y_3 + y_5 + y_7) + 2(y_2 + y_4 + y_6)\right]$$

$$= \frac{0.0625}{3}\left[\begin{array}{l}(1+1.0429) + 4\left(\begin{array}{l}1.0006 + 0.10059\\+1.0165 + 1.0326\end{array}\right)\\+2(1.0026+1.0105+1.1003)\end{array}\right]$$

$$= 0.510253$$

Using Romberg's formulae, we obtain

242

$$I\left(h, h/2\right) = \frac{1}{3}\left[4I\left(\frac{h}{2}\right) - I\left(h\right)\right] = 0.5241$$

$$I\left(\frac{h}{2}, \frac{h}{4}\right) = \frac{1}{3}\left[4I\left(\frac{h}{4}\right) - I\left(\frac{h}{2}\right)\right] = 0.5070$$

$$I\left(h, \frac{h}{2}, \frac{h}{4}\right) = \frac{1}{3}\left[4I\left(\frac{h}{2}, \frac{h}{4}\right) - I\left(h, \frac{h}{2}\right)\right] = 0.5013$$

$$Hence \int_{0}^{0.5} \frac{x}{\sin x}\, dx = 0.501$$

4.6. GAUSSIAN INTEGER

So far the formulae derived for evaluation of $\int_{a}^{b} f(x)\, dx$,

required the values of the function at equally spaced points of the interval. Gauss derived a formula which uses the same number of functional values but with different spacing and yields better accuracy. Gauss formula is expressed as

$$\int_{-1}^{1} f(x)\, dx = w_1 f(x_1) + w_2 f(x_2) + \dots + w_n f(x_n) = \sum_{i=1}^{n} w_i f(x_i)$$

where w_i and x_i are called the *weights* and *abscissae,* respectively. *The abscissae and weights are symmetrical with respect to the middle point of the interval.* There being $2n$ unknowns, $2n$ relations between them are necessary so that the formula is exact for all polynomials of degree not exceeding $2n - 1$. Thus we consider

$$f(x) = c_0 + c_1 x + c_2 x^2 + \dots + c_{2n-1} x^{2n-1}$$

243

Then

$$\int_{-1}^{1} f(x)\,dx = \int_{-1}^{1} \left(c_0 + c_1 x + c_2 x^2 + \ldots + c_{2n-1} x^{2n-1} \right) dx$$

$$= 2c_0 + \frac{2}{3} c_2 + \frac{2}{5} c_4 + \ldots$$

Putting $x = x_i$, we get

$$f(x_i) = c_0 + c_1 x_1 + c_2 x_1^2 + c_3 x_1^3 + \ldots + c_{2n-1} x_1^{2n-1}$$

Substituting these values on the right hand side, we obtain

$$\int_{-1}^{1} f(x)\,dx = w_1 \begin{pmatrix} c_0 + c_1 x_1 + c_2 x_1^2 + c_3 x_1^3 + \ldots + c_{2n-1} x_1^{2n-1} \\ + w_2 \left(c_0 + c_1 x_2 + c_2 x_2^2 + \ldots + c_{2n-1} x_2^{2n-1} \right) \\ + w_3 \left(c_0 + c_1 xn + c_2 xn^2 + c_3 xn^3 + \ldots + c_{2n-1} xn^{2n-1} \right) \\ + \ldots \end{pmatrix}$$

$$= c_0 \left(w_1 + w_2 + w_3 + \ldots + w_n \right) + c_1 \left(w_1 x_1 + w_2 x_2 + w_3 x_3 + \ldots + w_n x_n \right)$$

$$+ c_2 \left(w_1 x_1^2 + w_2 x_2^2 + \ldots + w_n x_n^2 \right) + \ldots$$

$$+ c_{2n-1} \left(w_1 x_1^{2n-1} + w_2 x_2^{2n-1} + w_3 x_3^{2n-1} + \ldots + w_n x_n^{2n-1} \right)$$

But the equations are identical for all values of c_i, hence comparing coefficients of c_i, we obtain 2n equations in 2n unknowns w_i and $x_i \left(i = 1, 2, ... n \right)$

$$w_1 + w_2 + w_3 + ... + w_n = 2$$

$$w_1 x_1 + w_2 x_2 + w_3 x_3 + ... + w_n x_n = 0$$

$$w_1 x_1^2 + w_2 x_2^2 + w_3 x_3^2 + ... + w_n x_n^2 = \frac{2}{3}$$

$$...$$

$$w_1 x_1^{2n-1} + w_2 x_2^{2n-1} + w_3 x_3^{2n-1} + ... + w_n x_n^{2n-1} = 0$$

The solution of the above equations is extremely complicated. It can however, be shown that x_i are the zeros of the (n+1)th legendre polynomial

Gauss formula for n=2 is

$$\int_{-1}^{1} f(x) dx = w_1 f(x_1) + w_2 f(x_2)$$

Then the equation becomes

$$w_1 + w_2 = 2$$

$$w_1 x_1 + w_2 x_2 = 0$$

$$w_1 x_1^2 + w_2 x_2^2 = \frac{2}{3}$$

$$w_1 x_1^3 + w_2 x_2^3 = 0$$

Solving these equations, we obtain

$$w_1 = w_2 = 1, x_1 -\frac{-1}{\sqrt{3}}, x_2 = \frac{1}{\sqrt{3}}$$

Thus Gauss formula for n=2 is

$$\int_{-1}^{1} f(x)dx = f\left(\frac{-1}{\sqrt{3}}\right) + f\left(\frac{1}{\sqrt{3}}\right)$$

Which gives the correct value of the integral of $f(x)$ in the range for any function up to third order. Equation is also known as Gauss Legendre formula. Gauss formula for n=3 is

$$\int_{-1}^{1} f(x)dx = \frac{8}{9} f(0) + \frac{5}{9}\left[f\left(-\sqrt{\frac{3}{5}}\right) + f\left(\sqrt{\frac{3}{5}}\right)\right]$$

Which is exact for polynomials up to degree 5.

The abscissae and the weights are tabulated for different values of n. The following table lists the abscissae and weights for values of n from 2 to 5.

Table 4.1. Gauss integration Abscissae and weights

N	x_i	w_i
2	-0.57735	1.0000
	0.57735	1.0000
3	-0.7746	0.5555
	0.0000	0.88889
	0.77460	0.55555
4	-0.86114	0.34785
	-0.33998	0.65214
	0.33998	0.65214
	0.86114	0.34785
5	-0.90618	0.23693
	-0.53847	0.47863
	0.0000	0.56889
	0.53847	0.47863
	0.90618	0.23693

Gauss formula imposes a restriction on the limits of integration to be from -1 to 1.

In general, the limits of the integral $\int_a^b f(x)\,dx$ are changed to -1 to 1 by means of the transformation

$$x = \frac{1}{2}(b-1)u + \frac{1}{2}(b+a)$$

Example.4.23 Evaluate $\int_{-1}^{1} \frac{1}{1+x^2}\,dx$ using Gauss formula for n=2 and n=3

Solution Gauss formula for n=2 is

$$I = \int_{-1}^{1} \frac{1}{1+x^2}\,dx = f\left(-\frac{1}{\sqrt{3}}\right) + f\left(\frac{1}{\sqrt{3}}\right) \text{ where } f(x) = \frac{1}{1+x^2}$$

$$\therefore I = \frac{1}{1+\left(-1/\sqrt{3}\right)^2} = \frac{3}{4} + \frac{3}{4} = 1.5$$

Gauss formula for n=3 is

$$I = \frac{8}{9}f(0) + \frac{5}{9}\left(\frac{5}{8} + \frac{5}{8}\right) = \frac{8}{9} + \frac{50}{72} = 1.5833$$

Example.4.24 Using the three-point Gaussian quadrature formula, evaluate $\int_0^1 \frac{dx}{1+x}$

Solution We first change the limits 0 and 1 to -1 to 1, so that

$$x = \frac{1}{2}(1-0)u\frac{1}{2}(1+0) = \frac{1}{2}(u+1)$$

$$\therefore I = \int_0^1 \frac{dx}{1+x} = \int_{-1}^1 \frac{1/2\,du}{1+1/2(u+1)} = \int_{-1}^1 \frac{du}{u+3}$$

Gauss-formula for n=3 is

$$I = \frac{8}{9}f(0) + \frac{5}{9}f\left(-\sqrt{\frac{3}{5}}\right) + f\left(\sqrt{\frac{3}{5}}\right) \text{ where } f(x) = \frac{1}{1+x^2}$$

Thus

$$t = \frac{8}{9}\left(\frac{1}{3}\right) + \frac{5}{9}\left\{\frac{1}{\sqrt{(3/5)+3}} + \frac{1}{\sqrt{(3/5)+3}}\right\}$$

$$= 0.6931$$

Example.4.25 Evaluate $\int_0^2 \frac{x^2+2x+1}{1+(x+1)^4}\,dx$ by the Gaussian three-point formula.

Solution Changing the limits of integration 0 to 2 to -1 to 1 by

$$x = \frac{1}{2}(b-a)u + (b+a) = \frac{2-0}{2}u + \frac{2+0}{2} = u+1$$

$$\therefore I = \int_0^2 \frac{x^2+2x+1}{1+(x+1)^4}\,dx = \int_{-1}^1 \frac{(u+1)^2 + 2(u+1)+1}{1+(u+1+u)^4}\,du$$

$$= \int_{-1}^{1} \frac{u^2 + 4u + 4}{(u+2)^4 + 1} du = \int_{-1}^{1} f(u) du$$

$$= w_1 f(u_1) + w_2 f(u_2) + w_3 f(u_3)$$

where $f(u_i) = \dfrac{u_i^2 + 4u_i + 4}{(u_i + 2)^4 + 1}$

Now

$$f(0) = \frac{4}{2^4 + 1} = \frac{4}{17}$$

$$f\left(-\sqrt{\frac{3}{5}}\right) = \frac{-\left(\sqrt{3/5} + 2\right)^2}{\left(-\sqrt{3/5} + 2\right)^4 + 1} = \frac{7.6984}{60.2652} = 0.1277$$

Using the three-point Gaussian formula, we have

$$I = \int_{-1}^{1} f(u) du = \frac{8}{9} f(0) + \frac{5}{9}\left[f\left(-\sqrt{\frac{3}{5}}\right) + f\left(\sqrt{\frac{3}{5}}\right)\right]$$

$$= \frac{8}{9}\left(\frac{4}{17}\right) + \frac{5}{9}[0.4614 + 0.1277] = 0.5365.$$

Exercise

1. Obtain an estimate of the number of subintervals that should br chosen so as guarantee that the error committed in evaluating $\int_{1}^{2} \dfrac{dx}{x}$ by trapezoidal rule is less than 0.001.

2. Evaluate $\int_{0}^{2} \dfrac{dx}{x^2 + 4}$ using the Romberg's method. Hence obtain an approximate value of pi.

3. Apply Romberg's method to evaluate $\int_{4}^{5.2} (\log x)\,dx$, given that

4.

X	4.0	4.2	4.4	4.6	4.8	5.0	5.2
Y	1.3863	1.4351	1.4816	1.5266	1.5686	1.6094	1.6486

5. Using the Gaussian two-point formula compute

   a. $\int_{-2}^{2} e^{x^2}\,dx$ b. $\sum_{1}^{n} x^3 = \left\{ \dfrac{n(n+1)}{2} \right\}^2$

6. Evaluate the following integrals, using the Gauss three-point formulae

   a. $\int_{2}^{4} (1+x^4)\,dx$ b. $\int_{5}^{3} \dfrac{4}{2x^2}\,dx$

7. Using the four point Gauss formula, compute $\int_{0}^{1} x\,dx$ correct to four decimal places.

Numerical solution of ordinary differential equations

5.1. Runge-kutta methods

These methods are named after two german mathematicians carl Runge and wilheln kutta. They were developed to avoid the computation of higher order derivatives in Taylor's method. The Runge-kutta methods are designed to give greater accuracy and they possess the advantage of requiring only the function values at some selected points on the subinterval. By Euler's method,

$$y_1 = y_0 + hf\left(x_0, y_0\right)$$
$$= y_0 + hy_0^1 \left(\because y^1 = f\left(x, y\right)\right)$$

The Taylor's series method of solving differential equations numerically is restricted by the labor involved in finding the higher order derivatives. However, there is a class of methods known as Runge-Kutta methods which do not require the calculations of higher order derivatives and give greater accuracy. The Runge-Kutta formulae possess the advantage of requiring only the function values at some selected points. These methods agree with Taylor's series solution up to the term in h^r where r differs from method to method and is called the order of that method.

5.1.1. First order R-K method.

We have seen that Euler's method (Section 10.4) gi

$$y_1 = y_0 + hf(x_0, y_0) = y_0 + hy_0' \qquad [\because y' = f(x, y)]$$

Expanding by Taylor's series

$$y_1 = y(x_0 + h) = y_0 + hy_0' + \frac{h^2}{2}y_0'' + \cdots$$

It follows that the Euler's method agrees with the Taylor's series solution upto the term in h.

Hence, Euler's method is the Runge-Kutta method of the first order. Second order R-K method. The modified Euler's method gives

$$y_1 = y + \frac{h}{2}[f(x_0, y_0) + f(x_0 + h, y_1)]$$

Substituting $y_1 = y_0 + hf(x_0, y_0)$ on the right-hand side of, we obtain

$$y_1 = y_0 + \frac{h}{2}[f_0 + f(x_0 + h), y_0 + hf_0]$$

where $\qquad f_0 = (x_0, y_0)$

Expanding L.H.S. by Taylor's series, we get

$$y_1 = y(x_0 + h) = y_0 + hy_0' + \frac{h^2}{2!}y_0'' + \frac{h^3}{3!}y_0''' + \cdots$$

Expanding, $f(x_0 + h, y_0 + hf_0)$ by Taylor's series for a function of two variables, gives

$$y_1 = y_0 + \frac{h}{2}\left[f_0 + f_0 = (x_0, y_0) + h\left(\frac{\partial f}{\partial x}\right)_{\{0} + hf_0\left(\frac{\partial f}{\partial y}\right)_{\{0}\right\} + O(h^2)$$

$$= y_0 + \frac{1}{2}\left[hf_0 + hf_0 + h^2\left(\frac{\partial f}{\partial x}\right)_{\{0} + \left(\frac{\partial f}{\partial y}\right)_{\{0}\right\} + O(h^3)$$

$$= y_0 + hf_0 + \frac{h^2}{2}f_0' + O(h^3)$$

$$\left[\because \frac{df(x,y)}{dx} = \frac{\partial f}{\partial x} + f\frac{\partial f}{\partial y}\right]$$

$$= y_0 + hy_0' + \frac{h^2}{2!}y_0'' + O(h^3)$$

It follows that the modified Euler's method and agrees with the Taylor's series solution upto the term in h^2.

Hence the modified Euler's method is the Runge-Kutta method of the second order.

The second order Runge-Kutta formula is

$$y_1 = y_0 + \frac{1}{2}(k_1 + k_2)$$

where $k_1 = hf(x_0, y_0)$ and $k_2 = hf(x_0 + h, y_0 + k)$

5.1.2. Second order Runge - kutta method

Consider the differential equation $\dfrac{dy}{dx} = f(x, y)$ with the initial-condition. $y(x_0) = y_0, \dfrac{dy}{dx} = f(x, y) \Rightarrow dy = f(x, y) dx$

Integrating the above equation over $[x_0, x_1]$ using the trapezoidal rule,

$$\int_{x_0}^{x_1} dy = \int_{x_0}^{x_1} f(x, y) dx$$

$$\Rightarrow (y)_{x_0}^{x_1} = \frac{h}{2}\left[f(x_0, y_0) + f(x_1, y_1) \right]$$

$$\Rightarrow y(x_1) - y(x_0) = \frac{h}{2}\left[f(x_0, y_0) + f(x_1, y_1) \right]$$

$$\Rightarrow y_1 = y_0 + \frac{h}{2}\left[f(x_0, y_0) + f(x_1, y_1) \right]$$

$$\Rightarrow y_1 = y_0 + \frac{h}{2}\Big[f\left(x_0, y_0\right) + f\left(x_0 + h, y_0 + hf\left(x_0, y_0\right)\right)\Big]$$

$$\because x_1 = x_0 + h, y_1 = y_0 + hf\left(x_0, y_0\right)$$

Now we put

$$k_1 = hf\left(x_0, y_0\right)$$
$$k_2 = hf\left(x_0 + h, y_0 + hf\left(x_0, y_0\right)\right)$$
$$= hf\left(x_0 + h, y_0 + k_1\right)$$

By using above equation,

$$y_1 = y_0 + \frac{1}{2}\big[k_1 + k_2\big]$$

Similarly for finding y_2,

$$y_2 = y_1 + \frac{1}{2}\left(k_1 + k_2\right)$$
$$k_1 = hf\left(x_1, y_1\right)$$
$$k_2 = hf\left(x_1 + h, y_1 + k_1\right)$$

This is the Runge -kutta second order formula.

5.1.3. Third order R-K method.

Similarly, it can be seen that Runge's method agrees with Taylor's series solution up to the term in h^3.

As such, Runge's method is the Runge-Kutta method of the third order.

The third-order Runge-Kutta formula is

254

$$y_1 = y_0 + \frac{1}{6}(k_1 + 4k_2 + k_3)$$

$$k_1 = hf(x_0, y_0), \quad k_2 = hf\left(x_0 + \frac{1}{2}h, y_0 + \frac{1}{2}k_1\right)$$

$$k_3 = hf(x_0 + h, y_0 + k'),$$

$$where\ k' = k_3 = hf(x_0 + h, y_0 + k_1).$$

5.1.4. Fourth order R-K method.

Fourth-order Runge-kutta method is commonly known as Runge - kutta method. For this, we need to find the following calculations.

$$k_1 = hf\left(x_0, y_0\right)$$

$$k_2 = hf\left(x_0 + \frac{h}{2}, y_0 + \frac{k_1}{2}\right)$$

$$k_3 = hf\left(x_0 + \frac{h}{2}, y_0 + k_3\right)$$

$$k_4 = hf\left(x_0 + h, y_0 + h\right)$$

$$k = \frac{1}{6}\left[k_1 + 2k_2 + 2k_3 + k_4\right]$$

$$y_1 = y_0 + k$$

Similarly, we can find $y_2, y_3, \ldots$

It will be noted that if $f(x, y)$ is independent of y then the method reduces to Simpson's formula. Runge-kuttta formulae do not require prior calculations of higher derivatives of $y(x)$. This method is very popular and extensively used. The error in Runge-kutta method is of order h^3.

Note. The third order Runge-kutta method is given by

$$y_1 = y_0 + \frac{1}{6}\left(k_1 + 4k_2 + k_3\right) \text{ where}$$

$$k_1 = hf\left(x_0, y_0\right)$$

$$k_2 = hf\left(x_0 + \frac{h}{2}, y_0 + \frac{k_1}{2}\right)$$

$$k_3 = hf\left(x_0 + h, y_0 + 2k_2 - k_1\right)$$

This method is most commonly used and is often referred to as the Runge-Kutta method only.

Working rule for finding the increment k of y corresponding to an increment h of x by Runge-Kutta method from

$$\frac{dy}{dx} = f(x, y), y(x_0)$$

Is as follows:

Calculate successively $k_1 = hf\left(x_0, y_0\right)$,

$$k_2 = hf\left(x_0 + \frac{1}{2}h, \ y_0 + \frac{1}{2}k_1\right),$$

$$k_3 = hf\left(x_0 + \frac{1}{2}h, \ y_0 + \frac{1}{2}k_2\right),$$

and $\qquad k_4 = hf(x_0 + h, \ y_0 + k_3),$

Finally compute $\qquad k = \frac{1}{6}(k_1 + 2k_2 + 2k_3 + k_4)$

which gives the required approximate value as $y_1 = y_0 + k$.

(Note that k is the weighted mean of $k_1, k_2, k_3,$ and k_4.

Observation. One of the advantages of these methods is that the operation is identical whether the differential equation is linear or non-linear.

Example.5.1. Using the Runge-Kutta method of fourth order, solve $\frac{dy}{dx} = \frac{y^2 - x^2}{y^2 + x^2}$ with $y(0) = 1$ at $x = 0.2, 0.4$.

Solution   We have $f(x, y) = \frac{y^2 - x^2}{y^2 + x^2}$

To find y (0.2)

Hence $\qquad\qquad x_0 = 0, y_0 = 1, h = 0.2$

$$k_1 = hf(x_0, y_0) = 0.2f(0,1) = 0.2000$$

$$k_2 = hf\left(x_0 + \frac{1}{2}h, \ y_0 + \frac{1}{2}k_1\right) = 0.2 \times$$
$f(0.1,1.1) = 0.19672$

$$k_3 = hf\left(x_0 + \frac{1}{2}h, \ y_0 + \frac{1}{2}k_2\right) = 0.2f(0.1,1.09836) = 0.1967$$

$$k_4 = hf(x_0 + h, \ y_0 + k_3) = 0.2f(0.2,1.1967) = 0.1891$$

$$k = \frac{1}{6}(k_1 + 2k_2 + 2k_3 + k_4)$$

$$= \frac{1}{6}[0.2 + 2(0.19672) + 2(0.1967) + 0.1891] = 0.19599$$

Hence $y(0.2) = y_0 + k = 1.196$.

To find $y(0.4)$:

Here $\quad x_1 = 0.2, \ y_1 = 1.196, \ h = 0.2$.

$$k_1 = hf(x_1, y_1) = 0.1891$$

$$k_2 = hf\left(x_1 + \frac{1}{2}h, y_1 + \frac{1}{2}k_1\right) = 0.2f(0.3,1.2906) = 0.1795$$

$$k_3 = hf\left(x_1 + \frac{1}{2}h, y_1 + \frac{1}{2}k_2\right) = 0.2f(0.3,1.2858) = 0.1793$$

$$k_4 = hf(x_1 + h, y_1 + k_3) = 0.2f(0.4, 1.3753) = 0.1688$$

$$k = \frac{1}{6}(k_1 + 2k_2 + 2k_3 + k_4)$$

$$= \frac{1}{6}[0.1891 + 2(0.1795) + 2(0.1793) + 0.1688] = 0.1792$$

Hence $y(0.4) = y_1 + k = 1.196 + 0.1792 = 1.3752.$

Example.5.2 Apply the Runge-Kutta method to find the approximate value of y for $x = 0.2$, in steps of *0.1*, if $\frac{dy}{dx} = x + y^2$, $y = 1$ where $x = 0$.

Solution   Given $(x, y) = x + y^2$.

Here we take h = 0.1 and carry out the calculations in two steps.

Step I. $x0 = 0,\ y0 = 1,\ h = 0.1$

$$\therefore\ \ k_1 = hf(x_0, y_0) = 0.1\,f(0,1) = 0.1000$$

$$k_2 = hf\left(x_0 + \frac{1}{2}h,\ y_0 + \frac{1}{2}k_1\right) = 0.1\,f(0.05, 1.1) = 0.1152$$

$$k_3 = hf\left(x_0 + \frac{1}{2}h,\ y_0 + \frac{1}{2}k_2\right) = 0.1\,f(0.05, 1.1152) = 0.1168$$

$$k_4 = hf(x_0 + h,\ y_0 + k_3) = 0.1\,f(0.1, 1.1168) = 0.1347$$

$$\therefore\ \ k = \frac{1}{6}(k_1 + 2k_2 + 2k_3 + k_4)$$

$$= \frac{1}{6}(0.1000 + 0.2304 + 0.2336 + 0.1347) = 0.1165$$

$$\text{Giving } y(0.1) = y_0 + k = 1.1165$$

Step II. $x_1 = x_0 + h = 0.1,\ y_1 = 1.1165,\ h = 0.1$

$$\therefore \ k_1 = hf(x_1, y_1) = 0.1f(0.1, 1.1165) = 0.1347$$

$$k_2 = hf\left(x_1 + \frac{1}{2}h, \ y_1 + \frac{1}{2}k_1\right) = 0.1f(0.15, 1.1838) = 0.1551$$

$$k_3 = hf\left(x_1 + \frac{1}{2}h, \ y_1 + \frac{1}{2}k_2\right) =$$
$$0.1f(0.15, 1.194) = 0.1576$$

$$k_4 = hf(x_1 + h, \ y_2 + k_3) = 0.1f(0.2, 1.1576) = 0.1823$$

$$\therefore \ k = \frac{1}{6}(k_1 + 2k_2 + 2k_3 + k_4) = 0.1571$$

$$\text{Hence } y(0.2) = y_1 + k = 1.2736$$

Example.5.3 Using the Runge-Kutta method of the fourth order, solve for y at $x = 1.2, \ 1.4$

from $\frac{dy}{dx} = \frac{2xy+e^x}{x^2+xe^x}$ given $x_0 = 1$, $y_0 = 0$

Solution We have $f(x, y) = \frac{2xy+e^x}{x^2+xe^x}$

To find $y(1.2)$

$$\text{Here } x_0 = 1, y_0 = 0, h = 0.2$$

$$k_1 = hf(x_0, y_0) = 0.2\frac{0 + e'}{1 + e'} = 0.1462$$

$$k_2 = hf\left(x_0 + \frac{h}{2}, y_0 + \frac{k_1}{2}\right) = 0.2\left\{\frac{2(1+0.1)(0+0.073)e^{1+0.1}}{(1+0.1)^2+(1+0.1)e^{1+0.1}}\right\}$$

$$= 0.1402$$

$$k_3 = hf\left(x_0 + \frac{1}{2}h, y_0 + \frac{1}{2}k_2\right)$$

$$= 0.2\left\{\frac{2(1 + 0.1)(0 + 0.07)e^{1.1}}{(1 + 0.1)^2 + (1 + 0.1)e^{1.1}}\right\}$$

$$=0.1399$$

$$k_4 = hf(x_0 + h, y_0 + k_3) = 0.2 \left\{ \frac{2(1.2)(0.1399)e^{1.2}}{(1.2)^2 + (1.2)e^{1.2}} \right\}$$

$$=0.1348$$

and $k = \frac{1}{6}(k_1 + 2k_2 + 2k_3 + k_4) = \frac{1}{6}[0.1462 + 0.2804 + 0.2798 + 0.1348] = 0.1402$

$$y(1.2) = y0 + k = 0 + 0.1402 = 0.1402.$$

To find y (1.4)

Here $x_1 = 1.2$, $y_1 = 0.1402$, $h = 0.2$

$$k_1 = hf(x_1, y_1) = 0.2f(1.2,0) = 0.1348$$

$$k_2 = hf\left(x_1 + \frac{h}{2}, y_1 + \frac{k_1}{2}\right) = 0.2f(1.3,0.2076) = 0.1303$$

$$k_3 = hf\left(x_1 + \frac{h}{2}, y_1 + \frac{k_2}{2}\right) = 0.2f(1.3,0.2053) = 0.1301$$

$$k_4 = hf(x_1 + h, y_1 + k_3) = 0.2f(1.3,0.2703) = 0.1260$$

$$\therefore \quad k = \frac{1}{6}(k_1 + 2k_2 + 2k_3 + k_4)$$

$$= \frac{1}{6}[0.1348 + 0.2606 + 0.2602 + 0.1260]$$

$$= 0.1303$$

Hence $y(1.4) = y_1 + k = 0.1402 + 0.1303 = 0.2705.$

Example.5.4 Use Runge-kutta method to approximate y when x=0.1 given that y=1. When x=0 and $\frac{dy}{dx} = x + y$.

Solution Here $f(x, y) = x + y$

$$x_0 = 0, y_0 = 1, h = 0.1$$

Using Runge-kutta method,

$$k_1 = hf(x_0, y_0)$$
$$= h(x_0 + y_0)$$
$$= 0.1(0+1)$$
$$= 0.1$$

$$k_2 = hf\left(x_0 + \frac{h}{2}, y_0 + \frac{k_1}{2}\right)$$
$$= h\left(x_0 + \frac{h}{2}, y_0 + \frac{k_1}{2}\right)$$
$$= 0.1\left(0 + \frac{0.1}{2}, 1 + \frac{0.1}{2}\right)$$
$$= 0.11$$

$$k_3 = hf\left(x_0 + \frac{h}{2}, y_0 + k_3\right)$$
$$= h\left(x_0 + \frac{h}{2} + y_0 + \frac{k_2}{2}\right)$$
$$= 0.1\left(0 + \frac{0.1}{2} + 1 + \frac{0.11}{2}\right)$$
$$= 0.1105$$

$$k_4 = hf(x_0 + h, y_0 + h)$$
$$= h(x_0 + h + y_0 + k_3)$$
$$= 0.1(0 + 0.1 + 1 + 0.1105)$$
$$= 0.12105$$

$$k = \frac{1}{6}\left[k_1 + 2k_2 + 2k_3 + k_4\right]$$

$$k = \frac{1}{6}\left[0.1 + 2(0.11) + 2(0.1105) + 0.12105\right]$$

$$k = 0.1103$$

$$y_1 = y_0 + k$$

$$y_1 = 1 + 0.1103$$

$$y_1 = 1.1103$$

$$y(0.1) = 1.1103$$

Example.5.5 Use Runge-kutta method to approximate y when

x=0.2 given that y=1 when x=0 and $\dfrac{dy}{dx} = x + y^2$

Solution   Here $f(x,y) = x + y^2$

$$x_0 = 0, y_0 = 1, h = 0.2$$

Using Runge-kutta method,

$$k_1 = hf(x_0, y_0)$$
$$= h(x_0 + y_0)$$
$$= 0.2(0 + 1)$$
$$= 0.2$$

$$k_2 = hf\left(x_0 + \frac{h}{2}, y_0 + \frac{k_1}{2}\right)$$
$$= h\left(x_0 + \frac{h}{2} + \left(y_0 + \frac{k_1}{2}\right)^2\right)$$

$$= 0.2 \left(0 + \frac{0.2}{2} + \left(1 + \frac{0.2}{2} \right)^2 \right)$$

$$= 0.262$$

$$k_3 = hf \left(x_0 + \frac{h}{2}, y_0 + \frac{k_2}{2} \right)$$

$$= h \left(x_0 + \frac{h}{2} + \left(y_0 + \frac{k_2}{2} \right)^2 \right)$$

$$= 0.1 \left(0 + \frac{0.2}{2} + \left(1 + \frac{0.262}{2} \right)^2 \right)$$

$$= 0.2758$$

$$k_4 = hf \left(x_0 + h, y_0 + h \right)$$

$$= h \left(x_0 + h + \left(y_0 + k_3 \right)^2 \right)$$

$$= 0.2 \left(0 + 0.2 + \left(1 + 0.2758 \right)^2 \right)$$

$$= 0.3655$$

$$k = \frac{1}{6} \left[k_1 + 2k_2 + 2k_3 + k_4 \right]$$

$$k = \frac{1}{6} \left[0.2 + 2 \left(0.262 \right) + 2 \left(0.2758 \right) + 0.3655 \right]$$

$$k = 0.2755$$

$$y_1 = y_0 + k$$

$$y_1 = 1 + 0.2755$$

$$y_1 = 1.2755$$

$$y(0.2) = 1.2755$$

Example.5.6 Use Runge-kutta fourth order method to find the value of y when x=1 given that y=1 when x=0, $\dfrac{dy}{dx} = \dfrac{y-x}{y+x}$

Solution Here $f(x,y) = \dfrac{y-x}{y+x}$

$$x_0 = 0, y_0 = 1, h = 1$$

Using Runge-Kutta method,

$$k_1 = hf(x_0, y_0)$$

$$= h\left(\frac{y_0 - x_0}{y_0 + x_0}\right)$$

$$= 1\left(\frac{1-0}{1+0}\right)$$

$$= 1$$

$$k_2 = hf\left(x_0 + \frac{h}{2}, y_0 + \frac{k_1}{2}\right)$$

$$= h\left(\frac{y_0 + \dfrac{k_1}{2} - \left(x_0 + \dfrac{h}{2}\right)}{y_0 + \dfrac{k_1}{2} + \left(x_0 + \dfrac{h}{2}\right)}\right)$$

$$= \frac{1 + \dfrac{1}{2} - \left(0 + \dfrac{1}{2}\right)}{1 + \dfrac{1}{2} + 0 + \dfrac{1}{2}}$$

$$k_2 = 0.5$$

$$k_3 = hf\left(x_0 + \frac{h}{2}, y_0 + \frac{k_2}{2}\right)$$

$$= h\left(\frac{y_0 + \frac{k_2}{2} - \left(x_0 + \frac{h}{2}\right)}{y_0 + \frac{k_2}{2} + \left(x_0 + \frac{h}{2}\right)}\right)$$

$$= \left(\frac{1 + \frac{0.5}{2} - \left(0 + \frac{1}{2}\right)}{1 + \frac{0.5}{2} + \left(0 + \frac{1}{2}\right)}\right)$$

$$= 0.428$$

$$k_4 = hf\left(x_0 + h, y_0 + h\right)$$

$$= h\left(\frac{y_0 + k_3 - \left(x_0 + h\right)}{y_0 + k_3 - \left(x_0 + h\right)}\right)$$

$$= 1\left(\frac{1 + 0.428 - \left(0 + 1\right)}{1 + 0.428 + \left(0 + 1\right)}\right)$$

$$= 0.176$$

$$k = \frac{1}{6}\left[k_1 + 2k_2 + 2k_3 + k_4\right]$$

$$k = \frac{1}{6}\left[1 + 2(0.5) + 2(0.428) + 0.176\right]$$

$$k = 0.505$$

$$y_1 = y_0 + k$$

$$y_1 = 1 + 0.505$$

$$y_1 = 1.505$$

$$y(1) = 1.505$$

Example.5.7 Use Runge- Kutta method to solve

$10\dfrac{dy}{dx} = x^2 + y^2, y(0) = 1$ for the interval $0 \le x \le 4$ with

h=0.2.

Solution   Here $f(x,y) = \dfrac{x^2 + y^2}{10}, x_0 = 0, y_0 = 1, h = 0.2$

By Runge- Kutta method,

To find y_1

$$k_1 = hf(x_0, y_0)$$
$$= \frac{h}{10}(x_0^2 + y_0^2)$$
$$= \frac{0.2}{10}(0+1)$$
$$= 0.02$$

$$k_2 = hf\left(x_0 + \frac{h}{2}, y_0 + \frac{k_1}{2}\right)$$
$$= \frac{0.2}{10}\left[\left(0 + \frac{0.2}{2}\right)^2 + \left(1 + \frac{0.02}{2}\right)^2\right]$$
$$= 0.0206$$
$$k_3 = hf\left(x_0 + \frac{h}{2}, y_0 + \frac{k_2}{2}\right)$$
$$= \frac{h}{10}\left(\left(x_0 + \frac{h}{2}\right)^2 + \left(y_0 + \frac{k_2}{2}\right)^2\right)$$

266

$$= \frac{0.2}{10}\left[\left(0+\frac{0.2}{2}\right)^2 + \left(1+\frac{0.0206}{2}\right)^2\right]$$

$$k_3 = 0.0206$$

$$k_4 = hf\left(x_0 + h, y_0 + h\right)$$
$$= \frac{h}{10}\left[\left(x_0 + h\right)^2 + \left(y_0 + k_3\right)^2\right]$$
$$= \frac{0.2}{10}\left[\left(0+0.2\right)^2 + \left(1+0.0206\right)^2\right]$$
$$= 0.0216$$

$$k = \frac{1}{6}\left[k_1 + 2k_2 + 2k_3 + k_4\right]$$
$$k = \frac{1}{6}\left[0.02 + 2\left(0.0206\right) + 2\left(0.0206\right) + 0.0216\right]$$
$$k = 0.0206$$
$$y_1 = y_0 + k$$
$$y_1 = 1 + 0.0206$$
$$y_1 = 1.0206$$
$$y(1) = y(0.2) = 1.0206$$

To find y_2

$$k_1 = hf\left(x_1, y_1\right)$$
$$= \frac{h}{10}\left(x_1^2 + y_1^2\right)$$
$$= \frac{0.2}{10}\left(0.2^2 + 1.0206^2\right)$$
$$= 0.0216$$

$$k_2 = hf\left(x_1 + \frac{h}{2}, y_1 + \frac{k_1}{2}\right)$$

$$= \frac{0.2}{10}\left[\left(0.2 + \frac{0.2}{2}\right)^2 + \left(1.0206 + \frac{0.0216}{2}\right)^2\right]$$

$$= 0.0230$$

$$k_3 = hf\left(x_1 + \frac{h}{2}, y_1 + \frac{k_2}{2}\right)$$

$$= \frac{h}{10}\left(\left(x_1 + \frac{h}{2}\right)^2 + \left(y_1 + \frac{k_2}{2}\right)^2\right)$$

$$= \frac{0.2}{10}\left[\left(0.2 + \frac{0.2}{2}\right)^2 + \left(1.0206 + \frac{0.023}{2}\right)^2\right]$$

$$k_3 = 0.0231$$

$$k_4 = hf\left(x_1 + h, y_1 + h\right)$$

$$= \frac{h}{10}\left[\left(x_1 + h\right)^2 + \left(y_1 + k_3\right)^2\right]$$

$$= \frac{0.2}{10}\left[\left(0.2 + 0.2\right)^2 + \left(1.0206 + 0.0231\right)^2\right]$$

$$= 0.0249$$

$$k = \frac{1}{6}\left[k_1 + 2k_2 + 2k_3 + k_4\right]$$

$$k = \frac{1}{6}\left[0.0216 + 2(0.023) + 2(0.0231) + 0.0249\right]$$

$$k = 0.0231$$

$$y_1 = y_0 + k$$

$$y_1 = 1.0206 + 0.0231 = 1.0437$$

$$y_1 = 1.0437$$

$$y(2) = y(0.4) = 1.0437$$

Example.5.8 Find y(0.2) for $y'=x-y^2$, $x_0=0,y_0=1$, with step length 0.1 using Runge-Kutta 4 method (1st order derivative)

Solution   Given $y'=x-y^2, y(0)=1, h=0.1, y(0.2)=?$

Forth order R-K method

$$k_1 = hf(x_0, y_0) = (0.1) f(0,1)$$
$$= (0.1) \cdot (-0.5) = -0.05$$

$$k_2 = hf(x_0 + h_2, y_0 + \frac{k_1}{2})$$
$$= (0.1) f(0.05, 0.975)$$
$$= (0.1) \cdot (-0.4625) = -0.04625$$

$$k_3 = hf(x_0 + h_2, y_0 + \frac{k_2}{2})$$
$$= (0.1) f(0.05, 0.97688)$$
$$= (0.1) \cdot (-0.46344) = -0.04634$$

$$k_4 = hf(x_0 + h, y_0 + k_3)$$
$$= (0.1) f(0.1, 0.95366)$$
$$= (0.1) \cdot (-0.42683) = -0.04268$$

$$y_1 = y_0 + \frac{1}{6}(k_1 + 2k_2 + 2k_3 + k_4)$$

$y_1 = 1 + 16[-0.05 + 2(-0.04625) + 2(-0.046$

$y_1 = 0.95369$

$y(0.1) = 0.95369$

Again taking (x_1, y_1) in place of (x_0, y_0) and repeat the process

$k_1 = hf(x_1, y_1)$
$= (0.1) f(0.1, 0.95369)$
$= (0.1) \cdot (-0.42684) = -0.04268$

$$k_2 = hf(x_1 + h_2, y_1 + \frac{k_1}{2})$$
$$= (0.1) f(0.15, 0.93235)$$
$$= (0.1) \cdot (-0.39117) = -0.03912$$

$$k_3 = hf(x_1 + h_2, y_1 + \frac{k_2}{2})$$
$$= (0.1) f(0.15, 0.93413)$$
$$= (0.1) \cdot (-0.39206) = -0.03921$$

$$k_4 = hf(x_1 + h, y_1 + k_3)$$
$$= (0.1) f(0.2, 0.91448)$$
$$= (0.1) \cdot (-0.35724) = -0.03572$$

$$y_1 = y_0 + \frac{1}{6}(k_1 + 2k_2 + 2k_3 + k_4)$$

270

$$y_2 = 0.95369 + 16 \begin{bmatrix} -0.04268 + 2(-0.03912) \\ +2(-0.03921) + (-0.03572) \end{bmatrix}$$

$$y_2 = 0.91451$$

$$\therefore y(0.2) = 0.91451$$

Example.5.9 Find y(0.5) for $y'=-2x-y$, $x0=0, y0=-1$, with step length 0.1 using Runge-Kutta 4 method (1st order derivative)

Solution   Given $y'=-2x-y, y(0)=-1, h=0.1, y(0.5)=?$

Fourth order R-K method,

$$k_1 = hf(x_0, y_0)$$
$$= (0.1) f(0, -1)$$
$$= (0.1) \cdot (1) = 0.1$$

$$k_2 = hf(x_0 + h_2, y_0 + \frac{k_1}{2})$$
$$= (0.1) f(0.05, -0.95)$$
$$= (0.1) \cdot (0.85) = 0.085$$

$$k_3 = hf(x_0 + h_2, y_0 + \frac{k_2}{2})$$
$$= (0.1) f(0.05, -0.9575)$$
$$= (0.1) \cdot (0.8575) = 0.08575$$

$$k_4 = hf(x_0 + h, y_0 + k_3)$$
$$= (0.1) f(0.1, -0.91425)$$
$$= (0.1) \cdot (0.71425) = 0.07142$$

$$y_1 = y_0 + \frac{1}{6}(k_1 + 2k_2 + 2k_3 + k_4)$$

$$y_1 = -1 + 16\left[0.1 + 2(0.085) + 2(0.08575) + (0.07142)\right]$$

$$y_1 = -0.91451$$

$$\therefore y(0.1) = -0.91451$$

Again taking (x_1, y_1) in place of (x_0, y_0) and repeat the process

$$k_1 = hf(x_1, y_1)$$
$$= (0.1) f(0.1, -0.91451)$$
$$= (0.1) \cdot (0.71451) = 0.07145$$
$$k_2 = hf(x_1 + h_2, y_1 + k_1/2)$$
$$= (0.1) f(0.15, -0.87879)$$
$$= (0.1) \cdot (0.57879) = 0.05788$$
$$k_3 = hf(x_1 + h_2, y_1 + k_2/2)$$
$$= (0.1) f(0.15, -0.88557)$$
$$= (0.1) \cdot (0.58557) = 0.05856$$
$$k_4 = hf(x_1 + h, y_1 + k_3)$$
$$= (0.1) f(0.2, -0.85596)$$
$$= (0.1) \cdot (0.45596) = 0.0456$$

$$y_2 = y_1 + \frac{1}{6}(k_1 + 2k_2 + 2k_3 + k_4)$$

$$y_2 = -0.91451 + 16\left[\begin{array}{c} 0.07145 + 2(0.05788) \\ +2(0.05856) + (0.0456) \end{array}\right]$$

$$y_2 = -0.85619$$

$$\therefore y(0.2) = -0.85619$$

Again taking (x_2, y_2) in place of (x_0, y_0) and repeat the process

272

$$k_1 = hf(x_2, y_2)$$
$$= (0.1) f(0.2, -0.85619)$$
$$= (0.1) \cdot (0.45619) = 0.04562$$

$$k_2 = hf(x_2 + h_2, y_2 + \frac{k_1}{2})$$
$$= (0.1) f(0.25, -0.83338)$$
$$= (0.1) \cdot (0.33338) = 0.03334$$

$$k_3 = hf(x_2 + h_2, y_2 + \frac{k_2}{2})$$
$$= (0.1) f(0.25, -0.83952)$$
$$= (0.1) \cdot (0.33952) = 0.03395$$

$$k_4 = hf(x_2 + h, y_2 + k_3)$$
$$= (0.1) f(0.3, -0.82224)$$
$$= (0.1) \cdot (0.22224) = 0.02222$$

$$y_3 = y_2 + \frac{1}{6}(k_1 + 2k_2 + 2k_3 + k_4)$$

$$y_3 = -0.85619 + 16 \begin{bmatrix} 0.04562 + 2(0.03334) \\ +2(0.03395) + (0.02222) \end{bmatrix}$$

$$y_3 = -0.82246$$
$$\therefore y(0.3) = -0.82246$$

Again taking (x_3, y_3) in place of (x_0, y_0) and repeat the process

$$k_1 = hf(x_3, y_3)$$
$$= (0.1) f(0.3, -0.82246)$$
$$= (0.1) \cdot (0.22246) = 0.02225$$

$$k_2 = hf(x_3 + h_2, y_3 + \frac{k_1}{2})$$
$$= (0.1) f(0.35, -0.81133)$$
$$= (0.1) \cdot (0.11133) = 0.01113$$

$$k_3 = hf(x_3 + h_2, y_3 + \frac{k_2}{2})$$
$$= (0.1) f(0.35, -0.81689)$$
$$= (0.1) \cdot (0.11689) = 0.01169$$
$$k_4 = hf(x_3 + h, y_3 + k_3)$$
$$= (0.1) f(0.4, -0.81077)$$
$$= (0.1) \cdot (0.01077) = 0.00108$$

$$y_4 = y_3 + \frac{1}{6}(k_1 + 2k_2 + 2k_3 + k_4)$$

$$y_4 = -0.82246 + 16 \left[\begin{array}{c} 0.02225 + 2(0.01113) \\ +2(0.01169) + (0.00108) \end{array} \right]$$

$$y_4 = -0.81096$$
$$\therefore y(0.4) = -0.81096$$

Again taking (x_4, y_4) in place of (x_0, y_0) and repeating the process

$$k_1 = hf(x_4, y_4)$$
$$= (0.1) f(0.4, -0.81096)$$
$$= (0.1) \cdot (0.01096) = 0.0011$$

$$k_2 = hf(x_4 + h_2, y_4 + \frac{k_1}{2})$$
$$= (0.1) f(0.45, -0.81041)$$
$$= (0.1) \cdot (-0.08959) = -0.00896$$

$$k_3 = hf(x_4 + h_2, y_4 + \frac{k_2}{2})$$
$$= (0.1) f(0.45, -0.81544)$$
$$= (0.1) \cdot (-0.08456) = -0.00846$$
$$k_4 = hf(x_4 + h, y_4 + k_3)$$
$$= (0.1) f(0.5, -0.81942)$$
$$= (0.1) \cdot (-0.18058) = -0.01806$$

$$y_5 = y_4 + 16(k_1 + 2k_2 + 2k_3 + k_4)$$
$$y_5 = -0.81096 + 16 \begin{bmatrix} 0.0011 + 2(-0.00896) \\ +2(-0.00846) + (-0.01806) \end{bmatrix}$$

$$y_5 = -0.81959$$
$$\therefore y(0.5) = -0.81959$$

Example.5.10 Find y(0.2) for y'=-y, x0=0,y0=1, with step length 0.1 using 4th order Runge-Kutta method.
Solution Given y'=-y, y(0)=1, h=0.1, y(0.2)=?
Fourth order R-K method

$$k_1 = hf(x_0, y_0)$$
$$= (0.1) f(0,1)$$
$$= (0.1) \cdot (-1) = -0.1$$

$$k_2 = hf(x_0 + h_2, y_0 + \frac{k_1}{2})$$

$$= (0.1) f(0.05, 0.95)$$

$$= (0.1) \cdot (-0.95) = -0.095$$

$$k_3 = hf(x_0 + h_2, y_0 + \frac{k_2}{2})$$

$$= (0.1) f(0.05, 0.9525)$$

$$= (0.1) \cdot (-0.9525) = -0.09525$$

$$k_4 = hf(x_0 + h, y_0 + k_3)$$

$$= (0.1) f(0.1, 0.90475)$$

$$= (0.1) \cdot (-0.90475) = -0.09048$$

$$y_1 = y_0 + 16(k_1 + 2k_2 + 2k_3 + k_4)$$

$$y_1 = 1 + 16 \begin{bmatrix} -0.1 + 2(-0.095) + \\ 2(-0.09525) + (-0.09048) \end{bmatrix}$$

$$y_1 = 0.90484$$

$$\therefore y(0.1) = 0.90484$$

Again taking (x_1, y_1) in place of (x_0, y_0) and repeat the process

$$k_1 = hf(x_1, y_1)$$
$$= (0.1) f (0.1, 0.90484)$$
$$= (0.1) \cdot (-0.90484) = -0.09048$$

$$k_2 = hf(x_1 + h_2, y_1 + \frac{k_1}{2})$$
$$= (0.1) f (0.15, 0.8596)$$
$$= (0.1) \cdot (-0.8596) = -0.08596$$

$$k_3 = hf(x_1 + h_2, y_1 + \frac{k_2}{2})$$
$$= (0.1) f (0.15, 0.86186)$$
$$= (0.1) \cdot (-0.86186) = -0.08619$$

$$k4 = hf(x1 + h, y1 + k3)$$
$$= (0.1) f (0.2, 0.81865)$$
$$= (0.1) \cdot (-0.81865) = -0.08187$$

$$y_2 = y_1 + 16(k_1 + 2k_2 + 2k_3 + k_4)$$

$$y_2 = 0.90484 + 16 \begin{bmatrix} -0.09048 + 2(-0.08596) \\ +2(-0.08619) + (-0.08187) \end{bmatrix}$$

$$y_2 = 0.81873$$
$$\therefore y(0.2) = 0.81873$$

Example.5.11 Find y(0.2) for $y'=x-y2$, $x0=0, y0=1$, with step length 0.1 using Runge-Kutta 4th order method .

Solution Given $y'=x-y2, y(0)=1, h=0.1, y(0.2)=?$

Fourth order R-K method

$$k_1 = f(x_0, y_0) = f(0,1) = -0.5$$

$$k_2 = f(x_0 + h_2, y_0 + h\frac{k_1}{2}) = f(0.05, 0.975) = -0.4625$$

$$k_3 = f(x_0 + h_2, y_0 + h\frac{k_2}{2}) = f(0.05, 0.9769) = -0.4634$$

$$k_4 = f(x_0 + h, y_0 + hk_3) = f(0.1, 0.9537) = -0.4268$$

$$y_1 = y_0 + \frac{h}{6}(k_1 + 2k_2 + 2k_3 + k_4)$$

$$y_1 = 1 + 0.16\left[-0.5 + 2(-0.4625) + 2(-0.4634) + (-0.4268)\right]$$

$$y_1 = 0.9537$$

$$\therefore y(0.1) = 0.9537$$

Again taking (x_1, y_1) in place of (x_0, y_0) and repeat the process.

$$k_1 = f(x_1, y_1) = f(0.1, 0.9537) = -0.4268$$

$$k_2 = f(x_1 + h_2, y_1 + h\frac{k_1}{2}) = f(0.15, 0.9323) = -0.3912$$

$$k_3 = f(x_1 + h_2, y_1 + h\frac{k_2}{2}) = f(0.15, 0.9341) = -0.3921$$

$$k_4 = f(x_1 + h, y_1 + hk_3) = f(0.2, 0.9145) = -0.3572$$

$$y_2 = y_1 + \frac{h}{6}(k_1 + 2k_2 + 2k_3 + k_4)$$

$$y_2 = 0.9537 + 0.16\left[\begin{array}{c} -0.4268 + 2(-0.3912) \\ +2(-0.3921) + (-0.3572) \end{array}\right]$$

$$y_2 = 0.9145$$

$$\therefore y(0.2) = 0.9145$$

Example.5.12 Find y(0.5) for $y'=-2x-y$, $x0=0$, $y0=-1$, with step length 0.1 using Runge-Kutta 4th order method.

Solution Given $y'=-2x-y$, $y(0)=-1$, $h=0.1$, $y(0.5)=?$

Fourth order R-K method,

$$k_1 = f(x_0, y_0) = f(0,-1) = 1$$

$$k_2 = f(x_0 + h_2, y_0 + h\frac{k_1}{2}) = f(0.05,-0.95) = 0.85$$

$$k_3 = f(x_0 + h_2, y_0 + h\frac{k_2}{2}) = f(0.05,-0.9575) = 0.8575$$

$$k_4 = f(x_0 + h, y_0 + hk_3) = f(0.1,-0.91425) = 0.71425$$

$$y_1 = y_0 + \frac{h}{6}(k_1 + 2k_2 + 2k_3 + k_4)$$

$$y_1 = -1 + 0.16[1 + 2(0.85) + 2(0.8575) + (0.71425)]$$

$$y_1 = -0.91451$$

$$\therefore y(0.1) = -0.91451$$

Again taking (x_1, y_1) in place of (x_0, y_0) and repeat the process

$$k_1 = f(x_1, y_1) = f(0.1,-0.91451) = 0.71451$$

$$k_2 = f(x_1 + h_2, y_1 + h\frac{k_1}{2}) = f(0.15,-0.87879) = 0.57879$$

$$k_3 = f(x_1 + h_2, y_1 + h\frac{k_2}{2}) = f(0.15,-0.88557) = 0.58557$$

$$k_4 = f(x_1 + h, y_1 + hk_3) = f(0.2,-0.85596) = 0.45596$$

$$y_2 = y_1 + \frac{h}{6}(k_1 + 2k_2 + 2k_3 + k_4)$$

$$y_2 = -0.91451 + 0.16\left[\begin{array}{c} 0.71451 + 2(0.57879) \\ +2(0.58557) + (0.45596) \end{array}\right]$$

$$y_2 = -0.85619$$

$$\therefore y(0.2) = -0.85619$$

Again taking (x_2, y_2) in place of (x_0, y_0) and repeat the process

$$k_1 = f(x_2, y_2) = f(0.2, -0.85619) = 0.45619$$

$$k_2 = f(x_2 + h_2, y_2 + h\frac{k_1}{2}) = f(0.25, -0.83338) = 0.33338$$

$$k_3 = f(x_2 + h_2, y_2 + h\frac{k_2}{2}) = f(0.25, -0.83952) = 0.33952$$

$$k_4 = f(x_2 + h, y_2 + hk_3) = f(0.3, -0.82224) = 0.22224$$

$$y_3 = y_2 + \frac{h}{6}(k_1 + 2k_2 + 2k_3 + k_4)$$

$$y_3 = -0.85619 + 0.16\left[\begin{array}{c} 0.45619 + 2(0.33338) \\ +2(0.33952) + (0.22224) \end{array}\right]$$

$$y_3 = -0.82246$$

$$\therefore y(0.3) = -0.82246$$

Again taking (x_3, y_3) in place of (x_0, y_0) and repeat the process.

$$k_1 = f(x_3, y_3) = f(0.3, -0.82246) = 0.22246$$

$$k_2 = f(x_3 + h_2, y_3 + h\frac{k_1}{2}) = f(0.35, -0.81133) = 0.11133$$

$$k_3 = f(x_3 + h_2, y_3 + h\frac{k_2}{2}) = f(0.35, -0.81689) = 0.11689$$

$$k_4 = f(x_3 + h, y_3 + hk_3) = f(0.4, -0.81077) = 0.01077$$

$$y_4 = y_3 + \frac{h}{6}(k_1 + 2k_2 + 2k_3 + k_4)$$

$$y_4 = -0.82246 + 0.16 \begin{bmatrix} 0.22246 + 2(0.11133) \\ +2(0.11689) + (0.01077) \end{bmatrix}$$

$$y_4 = -0.81096$$

$$\therefore y(0.4) = -0.81096$$

Again taking (x_4, y_4) in place of (x_0, y_0) and repeat the process.

$$k_1 = f(x_4, y_4) = f(0.4, -0.81096) = 0.01096$$

$$k_2 = f(x_4 + h_2, y_4 + h\frac{k_1}{2}) = f(0.45, -0.81041) = -0.08959$$

$$k_3 = f(x_4 + h_2, y_4 + h\frac{k_2}{2}) = f(0.45, -0.81544) = -0.08456$$

$$k_4 = f(x_4 + h, y_4 + hk_3) = f(0.5, -0.81942) = -0.18058$$

$$y_5 = y_4 + \frac{h}{6}(k_1 + 2k_2 + 2k_3 + k_4)$$

$$y_5 = -0.81096 + 0.16 \begin{bmatrix} 0.01096 + 2(-0.08959) \\ +2(-0.08456) + (-0.18058) \end{bmatrix}$$

$$y_5 = -0.81959$$

$$\therefore y(0.5) = -0.81959$$

281

Example.5.12 Find y(0.2) for $y'=-y$, $x0=0, y0=1$, with step length 0.1 using Runge-Kutta 4th order method.

Solution Given $y'=-y, y(0)=1, h=0.1, y(0.2)=?$

Fourth order R-K method,

$$k_1 = f(x_0, y_0) = f(0,1) = -1$$

$$k_2 = f(x_0 + h_2, y_0 + h\frac{k_1}{2}) = f(0.05, 0.95) = -0.95$$

$$k_3 = f(x_0 + h_2, y_0 + h\frac{k_2}{2}) = f(0.05, 0.9525) = -0.9525$$

$$k_4 = f(x_0 + h, y_0 + hk_3) = f(0.1, 0.90475) = -0.90475$$

$$y_1 = y_0 + \frac{h}{6}(k_1 + 2k_2 + 2k_3 + k_4)$$

$$y_1 = 1 + 0.16\left[-1 + 2(-0.95) + 2(-0.9525) + (-0.90475)\right]$$

$$y_1 = 0.90484$$

$$\therefore y(0.1) = 0.90484$$

Again taking (x_1, y_1) in place of (x_0, y_0) and repeat the process

$$k_1 = f(x_1, y_1) = f(0.1, 0.90484) = -0.90484$$

$$k_2 = f(x_1 + h_2, y_1 + h\frac{k_1}{2}) = f(0.15, 0.8596) = -0.8596$$

$$k_3 = f(x_1 + h_2, y_1 + h\frac{k_2}{2}) = f(0.15, 0.86186) = -0.86186$$

$$k_4 = f(x_1 + h, y_1 + hk_3) = f(0.2, 0.81865) = -0.81865$$

$$y_2 = y_1 + \frac{h}{6}(k_1 + 2k_2 + 2k_3 + k_4)$$

$$y_2 = 0.90484 + 0.16\begin{bmatrix} -0.90484 + 2(-0.8596) \\ +2(-0.86186) + (-0.81865) \end{bmatrix}$$

$$y_2 = 0.81873$$

$$\therefore y(0.2) = 0.81873$$

Exercises

1. Use Runge's method to approximate y when $x = 1.1$, given that $y = 1.2$ when $x = 1$ and $\frac{dy}{dx} = 3x + y^2$.

2. Using the Runge-Kutta method of order 4, find $y(0.2)$ given that $\frac{dy}{dx} = 3x + y^2, y(0) = 1$ taking $h = 0.1$.

3. Using the Runge-Kutta method of order 4, compute $y(0.2)$ and $y(0.4)$ from $10\frac{dy}{dx} = x^2 + y^2 \; y^{(0)} = 1$, taking $h = 0.1$.

4. Use the Runge Kutta method to find y when $x = 1.2$ in steps of 0.1, given that $\frac{dy}{dx} = x^2 + y^2 \; and \; y(1) = 1.5$.

5. Given $\frac{dy}{dx} = x^3 + y, y(0) = 2$.   Compute $y(0.2)$, $y(0.4)$, and $y(0.6)$ by the Runge-Kutta method of fourth order.

6. Find $y(0.1) \; and \; y(0.2)$ using the Runge-Kutta fourth order formula, given that $y' = x^2 - y \; and \; y(0) = 1$.

7. Using fourth order Runge-Kutta method, solve the following equation, taking each step of $h = 0.1$, given $y(0) = 3$. $\frac{dy}{dx}\left(\frac{4x}{y} - xy\right)$. Calculate y for $x = 0.1 \; and \; 0.2$.

8. Find by the Runge-Kutta method an approximate value of y for $x = 0.6$, given that $y = 0.41$ when $x = 0.4$ and $\frac{dy}{dx} = \sqrt{(x + y)}$.

9. Using the Runge-Kutta method of order 4, find $y(0.2)$ for the equation $\frac{dy}{dx} = \frac{y-x}{y+x}, y(0) = 1. \; Take \; h = 0.2$.

10. Using fourth order Runge-Kutta method, integrate $y' = -2x^3 + 12x^2 - 20x + 8.5$, using a step size of *0.5* and initial condition of $y = 1$ *at x = 0.*

11. Using the fourth order Runge-Kutta method, find *y at x = 0.1* given that $\frac{dy}{dx} = 3e^x + 2y$, $y(0) = 0$ *and* $h = 0.1$.

12. Given that $\frac{dy}{dx} = \frac{(y^2 - 2x)}{(y^2 + x)}$ and *y=1 at x=0,* find y for $x = 0.1$, 0.2, 0.3, 0.4, and 0.5.

5.2. Predictor-Corrector Methods:

If x_{i-1} and x_i are two consecutive mesh points, we have $x_i = x_{i-1} + h$. In Euler's method, we have

$$y_i = y_{i-1} + hf(x_0 + \overline{i-1}h, \, y_{i-1}); \; i = 1,2,3 \cdots$$

The modified Euler's method (Section 10.5), gives

$$y_i = y_{i-1} + \frac{h}{2}[f(x_{i-1}, \, y_{i-1}) + f(x_i, \, y_i)]$$

The value of y_i is first estimated by using the equation, then this value is inserted on the right side, giving a better approximation of y_i. This value of y_i is again substituted in to find a still better approximation of y_i . This step is repeated until two consecutive values of y_i agree. This technique of refining an initially crude estimate of y_i by means of a more accurate formula is known as predictor-corrector method. The equation is therefore called the predictor while serves as a corrector of y_i.

 In the methods so far described to solve a differential equation over an interval, only the value of *y* at the beginning of the interval was required. In the predictor-corrector methods, four prior values are needed for finding the value of *y* at x_i . Though slightly complex, these methods have the advantage of giving an estimate of error from successive approximations to y_i . We now describe two such methods, namely: Milne's method and Adams-Bash forth method.

Upto now we have discussed single- step methods and step- by step methods to solve the given ordinary differential equations of first order and first degree over a single interval (x_n, x_{n+1}). For this, we require the information only at $x = x_n$. Predictor – corrector methods require the function values $x_n, x_{n-1}, x_{n-2,......}$ for the computation of the function value of x_{n+1}.

A Predictor formula is used to predict the value of y_{n+1} and then a corrector formula is used to improve the value of y_{n+1}. Now we discuss Milne's method and Adam's –Bash forth –Moulton methods which are known as predictor-corrector methods. If a method requires four previous points y_0, y_1, y_2, y_3 then they can be calculated by single step methods discussed earlier.

5.3. Adams-bash forth-Moultan method

This is another predictor corrector method which uses backward differences Newton's backward difference interpolation formula can be written as

$$f(x,y) = f_0 + p\nabla f_0 + \frac{p(p+1)}{\angle 2}\nabla^2 f_0 + \frac{p(p+1)(p+2)}{\angle 3}\nabla^3 f_0 + ...$$

where $p = \dfrac{x - x_n}{h}, f_0 = f(x_0, y_0)$

We know that $y_1 = y_0 + \displaystyle\int_{x_0}^{x_1} f(x, y)$

$$\Rightarrow y_1 = y_0 + \int_{x_0}^{x_1} \left(f_0 + p\nabla f_0 + \frac{p(p+1)}{\angle 2}\nabla^2 f_0 + \frac{p(p+1)(p+2)}{\angle 3}\nabla^3 f_0 \right. \\ \left. + ... \right)$$

We have

$$p = \frac{x - x_n}{h} \Rightarrow dp = \frac{1}{h}dx$$
$$\Rightarrow dx = h.dp$$

Then

$$y_1 = y_0 + h\int_0^1 \left(f_0 + p\nabla f_0 + \frac{p(p+1)}{\angle 2}\nabla^2 f_0 + ... \right)$$

$$\Rightarrow y_1 = y_0 + h\left[f_0 + \frac{1}{2}\nabla f_0 + \frac{5}{12}\nabla^2 f_0 + \frac{3}{8}\nabla^3 f_0 + \frac{251}{720}\nabla^4 f_0 + ... \right]$$

By ignoring the higher order differences and expressing the lower order differences in terms of function values, we get

$$y_1 = y_0 + \frac{h}{24}\left[55 f_0 - 59 f_{-1} + 37 f_{-2} - 9 f_{-3} \right]$$

This is called Adams-Bash forth – predictor formula. Similarly, we can derive corrector formula by using Newton's backward difference formula at f_1.

$$f(x,y) = f_1 + p\nabla f_1 + \frac{p(p+1)}{\angle 2}\nabla^2 f_1 + \frac{p(p+1)(p+2)}{\angle 3}\nabla^3 f_1 + ...$$

Substituting the above formula in $y_1 = y_0 + \int_{x_0}^{x_1} f(x,y)$

286

$$\Rightarrow y_0 + h \int_0^1 \left(f_1 + p\nabla f_1 + \frac{p(p+1)}{\angle 2} \nabla^2 f_1 + \dots \right)$$

$$\Rightarrow y_0 + h \left(f_1 - \frac{1}{2}\nabla f_1 - \frac{1}{12}\nabla^2 f_1 - \frac{1}{24}\nabla^3 f_1 - \frac{19}{720}\nabla^4 f_1 \dots \right)$$

Thus by neglecting higher order differences, it can be written as

$$y_1 = y_0 + \frac{h}{24}[9f_1 + 19f_0 - 5f_{-1} + f_{-2}]$$

This is called Adam's-Moulton corrector formula.

Generally, these formulas can be written as

$$y_{n+1}^p = y_n + \frac{h}{24}[55f_n - 59f_{n-1} + 37f_{n-2} - 9f_{n-3}]$$

$$y_{n+1}^c = y_n + \frac{h}{24}\left[9f_{n+1}^p + 19f_n - 5f_{n-1} + f_{n-2}\right]$$

The super fixes p and c stand for predictor and corrector formula.

Procedure Consider the D.E. $\dfrac{dy}{dx} = f(x,y)$ with initial values

$$y(x_{-3}) = y_{-3}, y(x_{-2}) = y_{-2}, y(x_{-1}) = y_{-1}, y(x_0) = y_0$$

1. Find $f_0 = f(x_0, y_0), f_{-1} = f(x_{-1}, y_{-1}), f_{-2}$
   $$= f(x_{-2}, y_{-2}), f_{-3} = f(x_{-3}, y_{-3})$$

2. Using predictor formula, find
   $$y_1 = y_0 + \frac{h}{24}[55f_0 - 59f_{-1} + 37f_{-2} - 9f_{-3}]$$

3. Find $f_1 = f(x_1, y_1)$

4. Using corrector formula, find

$$y_1 = y_0 + \frac{h}{2}\left[9f_1 + 19f_0 - 5f_{-1} + f_{-2}\right]$$

Example. 5.14 Obtain the solution of the initial value problem $\frac{dy}{dx} = x^2 + x^2 y, y(1) = 1$ at x=1.4 by Adam's –Bash forth – Moulton method.

Solution Given $\frac{dy}{dx} = x^2 + x^2 y, y(1) = 1$, $f(x, y) = x^2 + x^2 y$

$y^1 = x^2 + x^2 y \Rightarrow y_0^1 = x_0^2 + x_0^2 y_0 = 2$

$y^{11} = 2x + x^2 y^1 + 2xy = y_0^{11} = 2x_0 + x_0^2 y_0^1 + 2x_0 y_0 = 6$

$y^{111} = 2 + x^2 y^{11} + 2xy^1 + 2xy^1 + 2y$

$= 2 + 4xy^1 + x^2 y^{11} + 2y \Rightarrow y_0^{11} = 2 + 4x_0 y_0^1 + x_0^2 y_0^{111} + 2y_0 = 18$

By Taylor's series method,

$$y = y_0 + (x - x_0) y_0^1 + \frac{(x - x_0)^2}{\angle 2} y_0^{11} + \frac{(x - x_0)^3}{\angle 3} y_0^{111} + \ldots$$

$$y = 1 + (x - 1) + \frac{(x - 1)^2}{\angle 2} 6 + \frac{(x - 1)^3}{\angle 3} 18 + \ldots$$

$$y(x) = 1 + 2(x - 1) + 3(x - 1)^2 + 3(x - 1)^3$$

$$x = 1.1, y(1.1) = y_{-2}$$

$$= 1 + 2(1.1 - 1) + 3(1.1 - 1)^2 + 3(1.1 - 1)^3 = 1.233$$

$$x = 1.2, y(1.2) = y_{-1}$$

$$= 1 + 2(1.2 - 1) + 3(1.2 - 1)^2 + 3(1.2 - 1)^3 = 1.544$$

$$x = 1.3, y(1.3) = y_0$$

$$= 1 + 2(1.3 - 1) + 3(1.3 - 1)^2 + 3(1.3 - 1)^3 = 1.951$$

288

i.e., $y_{-3} = 1, y_{-2} = 1.233, y_{-1} = 1.544, y_0 = 1.951$

$f_{-3} = f(x_{-3}, y_{-3}) = (x_{-3})^2 + (x_{-3})^2 y_{-3} = 2$

$f_{-2} = f(x_{-2}, y_{-2}) = (x_{-2})^2 + (x_{-2})^2 y_{-2} = 2.7019$

$f_{-1} = f(x_{-1}, y_{-1}) = (x_{-1})^2 + (x_{-1})^2 y_{-1} = 3.6633$

$f_0 = f(x_0, y_0) = (x_0)^2 + (x_0)^2 y_0 = 4.9871$

Using predictor formula,

$$y_1 = y_0 + \frac{h}{24}\left[55f_0 - 59f_{-1} + 37f_{-2} - 9f_{-3}\right]$$

$$= 1.951 + \frac{0.1}{24}\left[55(4.9871) - 59(3.6633) + 37(2.7019) - 9(2)\right]$$

$$= 2.5348$$

$$f_1 = f(x_1, y_1) = x_1^2 + x_1^2 y_1 = (1.4)^2 + (1.4)^2(2.5348)$$

$$= 6.9282$$

Using corrector formula,

$$y_1 = y_0 + \frac{h}{24}\left[9f_1 + 19f_0 - 5f_{-1} + f_{-2}\right]$$

$$= 1.951 + \frac{0.1}{24}\left[9(6.9282) + 19(4.9871) - 5(3.6633) + 2.7019\right]$$

$$= 2.5405$$

$$\therefore y(1.4) = 2.5405$$

Exercise

1. The differential equation $y' + y^2 = x^2$ is satisfied by

X	-0.2	0	0.2	0.4
Y	-1.00267	0	0.00267	0.02131

Using Adam's method, find the solution for x=0.6,0.8.

2. Solve $y' = x^2 y$ for x=0.8 given $y(0) = 1$ by Adam's Bash forth method.

3. Find $y(0.1), y(0.2), y(0.3)$ from $y' = x^2 - y$, $y(0) = 1$. Hence obtain $y(0.4)$ using Adams Bash forth method.

Example.5.15 Given $\dfrac{dy}{dx} = x^2(1+y)$ and

$y(1) = 1, y(1.1) = 1.233, y(1.2) = 1.548, y(1.3) = 1.979,$

evaluate $y(1.4)$ by the Adams-Bash forth method.

Solution Here $\dfrac{dy}{dx} = x^2(1+y)$

Starting values of the Adams-Bash forth method with h=0.1 are
$x = 1.0, y_{-3} = 1.000, f_{-3} = (1.0)^2 1 + 1.000 = 2.000$
$x = 1.1, y_{-2} = 1.233, f_{-2} = 2.702$
$x = 1.2, y_{-1} = 1.548, f_{-1} = 3.669$
$x = 1.3, y_0 = 1.979, f_{-1} = 5.035$

290

Using the corrector,

$$y_1^{(p)} = y_0 + \frac{h}{24}\left(55f_0 - 59f_{-1} - 37f_{-2} - 9f_{-3}\right)$$

$$x_4 = 1.4, y_1^{(p)} = 2.573, f_1 = 7.004$$

Using the corrector,

$$y_1^{(c)} = y_0 + \frac{h}{24}\left(9f_1 + 19f_0 - 5f_{-1} + 37f_{-2} - 9f_{-3}\right)$$

$$y_1^{(c)} = 1.979 + \frac{0.1}{24}\left(\begin{array}{l}9 + 7.004 + 19(5.035) \\ -5(3.669) + 2.702\end{array}\right)$$

$$= 2.575$$

Hence $y(1.4) = 2.575$

Example.5.16   Evaluate y'=x+y₂,

x_i	0	0.5	1	1.5
y_i	2	2.636	3.595	4.968

Find y(2) by Adams bash forth predictor method

Solution   Given $y'=x+y_2$

Adam's Bash forth Predictor formula is

$$y_{n+1}, p = y_n + \frac{h_2}{4}\left(55y_n' - 59y_n' - 1 + 37y_n' - 2 - 9y_n' - 3\right)$$

putting n=3, we get

$$y_4, p = y_3 + \frac{h_2}{4}(55y_3' - 59y_2' + 37y_1' - 9y_0')$$

We have given that
$$x_0 = 0, x_1 = 0.5, x_2 = 1, x_3 = 1.5$$

$$y_0 = 2, y_1 = 2.636, y_2 = 3.595, y_3 = 4.968,$$
$$y' = x + y_2 y_0' = x + y_2 = 1 \text{ (where } x = 0, y = 2),$$
$$y_1' = x + y_1' = 1.568 \text{ (where } x = 0.5, y = 2.636)$$

$$y_2' = x + y_2 = 2.2975 \text{ (where } x = 1, y = 3.595)$$
$$y_3' = x + y_2 = 3.234 \text{ (where } x = 1.5, y = 4.968)$$

putting the values in above equations, we get
$$y_4, p = y_3 + \frac{h_2}{4}(55y_3' - 59y_2' + 37y_1' - 9y_0')$$
$$y_4, p = 4.968 + 0.524 \cdot (55 \cdot 3.234 - 59 \cdot 2.2975$$
$$+ 37 \cdot 1.568 - 9 \cdot 1)$$
$$y_4, p = 4.968 + 0.524 \cdot \binom{177.87 - 135.5525}{+58.016 - 9}$$
$$y4, p = 4.968 + 0.524 \cdot (91.3335)$$
$$y_4, p = 4.968 + 1.9028$$
$$y_4, p = 6.8708$$

Therefore the predicted value is 6.8708

Now, we will correct it by corrector method to get the final value
$$y_4' = x + y_2 = 4.4354 \text{ (where } x = 2, y = 6.8708)$$

Adam's Bash forth Corrector formula is
$$y_{n+1}, c = y_n + \frac{h_2}{4}(9y_n' + 1 + 19y_n' - 5y_{n-1}' + y_{n-2}')$$

putting $n=3$, we get

$$y_4, c = y_3 + \frac{h_2}{4}(9y_4' + 19y_3' - 5y_2' + y_1')$$

$$y_4, c = 4.968 + 0.524 \cdot (9 \cdot 4.4354 + 19 \cdot 3.234$$
$$- 5 \cdot 2.2975 + 1.568)$$

$$y_4, c = 4.968 + 0.524 \cdot \begin{pmatrix} 39.9185 + 61.446 \\ -11.4875 + 1.568 \end{pmatrix}$$

$$y_4, c = 4.968 + 0.524 \cdot (91.445)$$

$$y_4, c = 4.968 + 1.9051$$

$$y_4, c = 6.8731$$

$$\therefore y(2) = 6.8731$$

Example.5.17 If $y'=y\text{-}x_3$, $y(0)=1$. Find y(0.4) by Adams bash forth predictor method Step value (h) = 0.1. Initial values by Runge-Kutta 4th order method.

Solution Given $y'=y\text{-}x_3$

Adam's Bash forth Predictor formula ,

$$y_{n+1}, p = y_n + \frac{h_2}{4}(55y_n' - 59y_n' - 1 + 37y_n' - 2 - 9y_n' - 3)$$

Putting $n=3$, we get

$$y_4, p = y_3 + \frac{h_2}{4}(55y_3' - 59y_2' + 37y_1' - 9y_0')$$

We have given that

$$x_0 = 0, x_1 = 0.1, x_2 = 0.2, x_3 = 0.3$$

Initial values using Runge-Kutta 4th order method, we get

$y_0 = 1, y_1 = 1.1051, y_2 = 1.221, y_3 = 1.3477$

$y' = y - x_3 y'_0 = y - x_3 = 1$ (where $x = 0, y = 1$)

$y'_1 = y - x_3 = 1.1041$ (where $x = 0.1, y = 1.1051$)

$y'_2 = y - x_3 = 1.213$ (where $x = 0.2, y = 1.221$)

$y'_3 = y - x_3 = 1.3207$ (where $x = 0.3, y = 1.3477$)

Iteration-1 (for $x_4 = 0.4$)

$y_4, p = y_3 + \dfrac{h}{24}(55y'_3 - 59y'_2 + 37y'_1 - 9y'_0)$

$y_4, p = 1.3477 + 0.124 \cdot (55 \cdot 1.3207 - 59 \cdot 1.213$
$+ 37 \cdot 1.1041 - 9 \cdot 1)$

$y_4, p = 1.3477 + 0.124 \cdot \begin{pmatrix} 72.6388 - 71.5662 \\ +40.8534 - 9 \end{pmatrix}$

$y_4, p = 1.3477 + 0.124 \cdot (32.926)$

$y_4, p = 1.3477 + 0.1372$

$y_4, p = 1.4849$

So, the predicted value is 1.4849

Now, we will correct it by corrector method to get the final value

$y'4 = y - x3 = 1.4209$ (where $x = 0.4, y = 1.4849$)

Adam's Bash forth Corrector formula is

$y_{n+1}, p = y_n + \dfrac{h_2}{4}(55y'_n - 59y'_n - 1 + 37y'_n - 2 - 9y'_n - 3)$

putting $n=3$, we get

$$y_4, p = y_3 + \frac{h_2}{4}(55y_3' - 59y_2' + 37y_1' - 9y_0')$$

$$y_4, c = 1.3477 + 0.124 \cdot (9 \cdot 1.4209 + 19 \cdot 1.3207$$
$$- 5 \cdot 1.213 + 1.1041)$$

$$y_4, c = 1.3477 + 0.124 \cdot \begin{pmatrix} 12.7881 + 25.0934 \\ -6.0649 + 1.1041 \end{pmatrix}$$

$$y_4, c = 1.3477 + 0.124 \cdot (32.9207)$$

$$y_4, c = 1.3477 + 0.1372$$

$$y_4, c = 1.4849$$

$$\therefore y(0.4) = 1.4849$$

Example.5.18 *If $y'=x-y2$, $y(0)=1$.* Find y(1.0) by Adams bash forth predictor method Step value (h) = 0.2, initial values by Runge-Kutta 4th order method.

Solution   Given $y'=x-y_2$

By using Adams Bash forth Corrector formula is

$$y_{n+1}, p = y_n + \frac{h_2}{4}(55y_n' - 59y_n' - 1 + 37y_n' - 2 - 9y_n' - 3)$$

Putting $n=3$, we get

$$y_4, p = y_3 + \frac{h_2}{4}(55y_3' - 59y_2' + 37y_1' - 9y_0')$$

we have given that

$$x_0 = 0, x_1 = 0.2, x_2 = 0.4, x_3 = 0.6$$

Initial values using Runge-Kutta 4th order method, we get

$y_0 = 1, y_1 = 0.8512, y_2 = 0.7798, y_3 = 0.7621$

$y' = x - y_2 y'_0 = x - y_2 = -1$ (where $x = 0, y = 1$)

$y'_1 = x - y_2 = -0.5245$ (where $x = 0.2, y = 0.8512$)

$y'_2 = x - y_2 = -0.2081$ (where $x = 0.4, y = 0.7798$)

$y'_3 = x - y_2 = 0.0192$ (where $x = 0.6, y = 0.7621$)

Iteration-1 (for $x_4 = 0.8$)

$$y_4, p = y_3 + \frac{h}{24}(55y'_3 - 59y'_2 + 37y'_1 - 9y'_0)$$

$$y_4, p = 0.7621 + 0.224 \cdot (55 \cdot 0.0192 - 59 \cdot (-0.2081)$$
$$+ 37 \cdot (-0.5245) - 9 \cdot (-1))$$

$$y_4, p = 0.7621 + 0.224 \cdot (1.0559 + 12.2786 - 19.4077 + 9)$$

$$y_4, p = 0.7621 + 0.224 \cdot (2.9267)$$

$$y_4, p = 0.7621 + 0.0244$$

$$y_4, p = 0.7865$$

So, the predicted value is 0.7865

Now, we will correct it by corrector method to get the final value

$y'_4 = x - y_2 = 0.1814$ (where $x = 0.8, y = 0.7865$)

By using Adams Bash forth Corrector formula is

$$y_{n+1}, p = y_n + \frac{h_2}{4}(55y'_n - 59y'_n - 1 + 37y'_n - 2 - 9y'_n - 3)$$

Putting $n=3$, we get

$$y_4, p = y_3 + \frac{h_2}{4}(55y'_3 - 59y'_2 + 37y'_1 - 9y'_0)$$

$$y_4, c = 0.7621 + 0.224 \cdot (9 \cdot 0.1814 + 19 \cdot 0.0192$$
$$- 5 \cdot (-0.2081) + (-0.5245))$$

$$y_4, c = 0.7621 + 0.224 \cdot \begin{pmatrix} 1.6329 + 0.3648 \\ +1.0406 - 0.5245 \end{pmatrix}$$

$$y_4, c = 0.7621 + 0.224 \cdot (2.5136)$$

$$y_4, c = 0.7621 + 0.0209$$

$$y_4, c = 0.7831$$

$$y'_4 = x - y_2 = 0.1868 \text{ (where } x = 0.8, y = 0.7831)$$

$$y4, c = y3 + h24(9y'4 + 19y'3 - 5y'2 + y'1)$$
$$y4, c = 0.7621 + 0.224 \cdot (9 \cdot 0.1868 + 19 \cdot 0.0192$$
$$- 5 \cdot (-0.2081) + (-0.5245))$$
$$y4, c = 0.7621 + 0.224 \cdot (1.6815 + 0.3648 + 1.0406 - 0.5245)$$
$$y4, c = 0.7621 + 0.224 \cdot (2.5623)$$
$$y4, c = 0.7621 + 0.0214$$
$$y4, c = 0.7835$$

Iteration -2 (for $x_5 = 1$)

$$y_5, p = y4 + \frac{h}{24} (55y'_4 - 59y'_3 + 37y'_2 - 9y'_1)$$

$$y_5, p = 0.7835 + 0.224 \cdot (55 \cdot 0.1868 - 59 \cdot 0.0192$$
$$+ 37 \cdot (-0.2081) - 9 \cdot (-0.5245))$$

$$y_5, p = 0.7835 + 0.224 \cdot \begin{pmatrix} 10.2757 - 1.1327 \\ -7.7001 + 4.7208 \end{pmatrix}$$

$$y_5, p = 0.7835 + 0.224 \cdot (6.1637)$$

$$y_5, p = 0.7835 + 0.0514$$

$$y_5, p = 0.8348$$

$$y_{n+1}, c = y_n + \frac{h}{24}(9y'_{n+1} + 19y'_n - 5y'_{n-1} + y'_{n-2})$$

putting $n=4$, we get

$$y_5, c = y_4 + \frac{h}{24}(9y'_5 + 19y'_4 - 5y'_3 + y'_2)$$

$$y_5, c = 0.7835 + 0.224 \cdot (9(0.3031) + 19(0.1868)$$
$$- 5(0.0192) + (-0.2081))$$

$$y_5, c = 0.7835 + 0.224 \cdot \begin{pmatrix} 2.7277 + 3.5498 \\ -0.096 - 0.2081 \end{pmatrix}$$

$$y_5, c = 0.7835 + 0.224 \cdot (5.9734)$$

$$y_5, c = 0.7835 + 0.0498$$

$$y_5, c = 0.8332$$

$$y'_5 = x - y_2 = 0.3057 \text{ (where } x = 1, y = 0.8332)$$

$$y_5, c = y_4 + \frac{h}{24}(9y'_5 + 19y'_4 - 5y'_3 + y'_2)$$

$$y_5, c = 0.7835 + 0.224 \cdot (9(0.3057) + 19(0.1868)$$
$$- 5(0.0192) + (-0.2081))$$

$$y_5, c = 0.7835 + 0.224 \cdot \begin{pmatrix} 2.7515 + 3.5498 \\ -0.096 - 0.2081 \end{pmatrix}$$

$$y_5, c = 0.7835 + 0.224 \cdot (5.9972)$$

$$y_5, c = 0.7835 + 0.05$$

$$y_5, c = 0.8334$$

$$\therefore y(1) = 0.8334$$

Example.5.19   Given $\frac{dy}{dx} = x^2(1+y)$ and $y(1) = 1$,
$y(1.1) = 1.233$, $y(1.2) = 1.548$, $y(1.3) = 1.979$. Evaluate
$y(1.4)$ by the Adams-Bashforth method.

Solution Here $f(x,y) = x^2(1+y)$

Starting values of the Adams-Bash forth method with $h = 0.1$
are $x = 1.0$, $y_{-3} = 1.000$, $f_{-3} = (1.0)^2(1+1.000) =$
2.000

$$x = 1.1, \qquad y_{-2} = 1.233, \qquad f_{-2} = 2.702$$

$$x = 1.2, \qquad y_{-1} = 1.548, \qquad f_{-1} = 3.669$$

$$x = 1.3, \qquad y_0 = 1.979, \qquad f_0 = 5.035$$

Using the **predictor,**

$$y_1^{(p)} = y_0 + \frac{h}{24}(55f_0 - 59f_{-1} + 37f_{-2} - 9f_{-3})$$

$$x_4 = 1.4, \ y_1^{(p)} = 2.573 \qquad f_1 = 7.004$$

Using the **corrector,**

$$y_1^{(c)} = y_0 + \frac{h}{24}(9f_1 + 19f_0 - 5f_{-1} + f_{-2})$$

$$y_1^{(c)} = 1.979 + \frac{0.1}{24}(9 \times 7.004 + 19 \times 5.035 - 5 \times$$
$$3.669 + 2.702) = 2.575$$

Hence $y(1.4) = 2.575$

Example.5.20   If $\frac{dy}{dx} = 2e^x y, y(0) = 2$, find $y(4)$ using the
Adams predictor corrector formula by calculating $y(1), y(2)$,
and $y(3)$ using Euler's modified formula.

Solution We have $f(x,y) = 2e^x y$

x	$2e^x y$	Mean slope	Old y $+ h(\text{mean slop})$ $= \text{new } y$
0	4		$2 + 0.1(4) = 2.4$
0.1	$2e^{0.1}(2.4)$ $= 5.305$	$\frac{1}{2}(4 + 5.305)$ $= 4.6524$	$2 + 0.1(4.6524)$ $= 2.465$
0.1	$2e^{0.1}(2.465)$ $= 5.449$	$\frac{1}{2}(4 + 5.465)$ $= 4.7244$	$2 + 0.1(4.7244)$ $= 2.472$
0.1	$2e^{0.1}(2.4724)$ $= 5.465$	$\frac{1}{2}(4 + 5.465)$ $= 4.7324$	$2 + 0.1(4.7324)$ $= 2.473$
0.1	$2e^{0.1}(2.478)$ $= 5.467$	$\frac{1}{2}(4 + 5.467)$ $= 4.7333$	$2 + 0.1(4.7333)$ $= 2.473$
0.1	5.467	–	$2 + 0.1(5.467)$ $= 3.0199$
0.2	$2e^{0.1}(3.0199)$ $= 7.377$	$\frac{1}{2}(5.467$ $+ 7.377)$ $= 6.422$	2.473 $+ 0.1(6.422)$ $= 3.1155$
0.2	7.611	$\frac{1}{2}(5.467$ $+ 7.611)$ $= 6.539$	2.473 $+ 0.1(6.539)$ $= 3.127$
0.2	7.639	$\frac{1}{2}(5.467$ $+ 7.639)$ $= 6.553$	2.473 $+ 0.1(6.553)$ $= 3.129$

0.2	7.643	$\frac{1}{2}(5.467$ $+ 7.643)$ $= 6.555$	2.473 $+ 0.1(6.555)$ $= 3.129$
0.2	7.463	—	3.129 $+ 0.1(7.643)$ $= 3.893$
0.3	$2e^{0.3}(3.893)$ $= 10.51$	$\frac{1}{2}(7.643$ $+ 10.51)$ $= 9.076$	3.129 $+ 0.1(9.076)$ $= 4.036$
0.3	10.897	$\frac{1}{2}(7.643$ $+ 10.897)$ $= 9.266$	3.129 $+ 0.1(9.2696)$ $= 4.056$
0.3	10.949	$\frac{1}{2}(7.643$ $+ 10.949)$ $= 9.296$	3.129 $+ 0.1(9.296)$ $= 4.058$
0.3	10.956	$\frac{1}{2}(7.643$ $+ 10.956)$ $= 9.299$	3.129 $+ 0.1(9.299)$ $= 4.0586$

To find $y(0.4)$ by **Adam's method,** the starting values with $h = 0.1$ are

$$x = 0.0, \quad y_{-3} = 2.4, \qquad f_{-3} = 4$$

$$x = 0.1, \quad y_{-2} = 2.473, \qquad f_{-2} = 5.467$$

$$x = 0.2, \quad y_{-1} = 3.129, \qquad f_{-1} = 7.643$$

$$x = 0.3, \quad y_0 = 4.059, \qquad f_0 = 10.956$$

Using the predictor formula

$$y_1^{(p)} = y_0 + \frac{h}{24}(55f_0 - 59f_{-1} + 37f_{-2} - 9f_{-3})$$

$$= 4.059 + \frac{0.1}{24}(55 \times 10.957 - 59 \times 7.643 + 37 \times 5.467 - 9 \times 4)$$

$$= 5.383$$

$$x = 0.3 \quad y_1 = 5.383 \quad f_1 = 2e^{0.4}(5.383) = 16.061$$

Using the corrector formula,

$$y_1^{(c)} = y_0 + \frac{h}{24}(9f_1 + 19f_0 - 5f_{-1} + f_{-2})$$

$$= 4.0586 + \frac{0.1}{24}(9 \times 16.061 + 19 \times 10.956 - 5 \times 7.643 + 5.467)$$

$$= 5.392$$

Hence $y(0.4) = 5.392$

Example.5.21   Solve   the   initial   value   problem $\frac{dy}{dx} = x - y^2, y(0) = 1$ to find $y(0.4)$ by Adam's method. Starting solutions required are to be obtained using the Runge-Kutta method of the fourth order using step value $h = 0.1$.

Solution   We have $f(x, y) = x - y^2$.

To find $y(0.1)$

$$\text{Here } x_0 = 0, y_0 = 1, h = 0.1$$

$$\therefore \ k_1 = hf(x_0, y_0) = (0.1)f(0.1) = -0.1000$$

$$k_2 = hf\left(x_0 + \frac{1}{2}h, y_0 + \frac{1}{2}k_1\right)$$

$$= (0.1)f(0.05,0.95) = -0.08525$$

$$k_3 = hf\left(x_0 + \frac{1}{2}h, y_0 + \frac{1}{2}k_2\right)$$

$$= (0.1)f(0.05,0.9574) = -0.0867$$

$$k_4 = hf(x_0 + h, y_0 + k_3)$$

$$= (0.1)f(0.1,0.9137) = -0.07341$$

$$k = \frac{1}{6}(k_1 + 2k_2 + 2k_3 + k_4)0 = -0.0883$$

Thus $y(0.1) = y_1 = y_0 + k = 1 - 0.0883 = 0.9117$

To find $y(0.2)$

Here $x_1 = 0.1, \ y_1 = 0.9117, h = 0.1$

$$k_1 = hf(x_1, y_1) = (0.1) \times f(0.1,0.9117) = -0.0731$$

$$k_2 = hf\left(x_1 + \frac{1}{2}h, y_1 + \frac{1}{2}k_1\right) = (0.1)f(0.15,0.8751) = 0.061$$
$$k_4 = hf(x_1 + h, y_1 + k_3) = (0.1)f(0.2,0.8491) = 0.0521$$

$$k = \frac{1}{6}(k_1 + 2k_2 + 2k_3 + k_4)0 = 0.8494.$$

Thus $y(0.2) = y_2 = y_1 + k = 0.8494.$

To find $y(0.3)$

Here $x_2 = 0.2 , y_2 = 0.8494, \ h = 0.1.$

$$k_1 = hf(x_2, y_2) = (0.1) \times f(0.25, 0.8494) = 0.05$$

$$k_2 = hf\left(x_2 + \frac{1}{2}h, y_2 + \frac{1}{2}k_1\right) = (0.1) f(0.25, 0.8233) = 0.0428$$

$$k_3 = hf\left(x_2 + \frac{1}{2}h, y_2 + \frac{1}{2}k_2\right) = (0.1)f(0.25, 0.828) = 0.0436$$

$$k_4 = hf(x_2 + h, y_2 + k_3) = (0.1)f(0.3, 0.058) = 0.0349$$

$$k = \frac{1}{6}(k_1 + 2k_2 + 2k_3 + k_4)0 = 0.0438$$

Thus $y(0.3) = y_3 = y_2 + k = 0.8061.$

Now the starting values for the Milne's method are:

$$x_0 = 0.0, \quad y_0 = 1.0000, \quad f_0 = 0.0 - (0.1)^2 = 1.0000$$

$$x_1 = 0.1, \; y_1 = 0.9117, \; f_1 = 0.1 - (0.9117)^2 = -0.7312$$

$$x_2 = 0.2, \; y_2 = 0.8494, \; f_2 = 0.2 - (0.8494)^2 = -0.5215$$

$$x_3 = 0.3, \; y_3 = 0.8061, \; f_3 = 0.3 - (0.8061)^2 = -0.3498$$

Using the predictor,

$$y_1^{(p)} = y_0 + \frac{h}{24}(55f_0 - 59f_{-1} + 37f_{-2} - 9f_{-3})$$

For $x = 0.4$

$$= 0.8061 + \frac{0.1}{24}(55(-0.3498) - 59(-0.5215) +$$
$$37(-0.7312) - 9(-1)) = 0.7789$$

$$f_1 = -0.267$$

Using the corrector,

$$y_1^{(c)} = y_0 + \frac{h}{24}(9f_1 + 19f_0 - 5f_{-1} + f_{-2})$$

$$= 0.8061 + \frac{0.1}{24}(9(-0.2067) + 19(-0.398) -$$
$$5(-0.5215) - 0.7312)$$

$$= 0.7785$$

Hence $y(0.4) = 0.7785$

Exercise

1. Using the Adams-Bashforth method, obtain the solution of $\frac{dy}{dx} = x - y^2$ at $x = 0.8$, given the values

x:	0	0.2	0.4	0.6
y:	0	0.0200	0.0795	0.1762

2. Using the Adams-Bashforth formulae, determine $y(0.4)$ given the differential equation $\frac{dy}{dx} = \frac{1}{2}xy$ and the data:

x:	0	0.1	0.2	0.3
y:	1	1.002	1.0101	1.0228

3. Given $y' = x^2 - y, y(0) = 1$ and the starting values $y(0.1) = 0.90516$, $y(0.2) = 0.82127, y(0.3) = 0.74918$, evaluate $y(0.4)$ using the Adams Bashforth method.

4. Using the Adams-Bashforth method, find $y(4.4)$ given $5xy' + y^2 = 2, y(4) = 1, y(4,1) = 1.0049, y(4,2) = 1.0097$, and $y(4.3) = 10143$.

5. Given the differential equation $\frac{dy}{dx} = x^2y + x^2$ and the data:

x:	1	1.1	1.2	1.3
y:	1	1.233	1.548488	1.978921

Determine $y(1.4)$ by any numerical method.

6. Using the Adams-Bashforth method, evaluate $y(1.4)$; if y satisfies $\frac{dy}{dx} + \frac{y}{x} = \frac{1}{x^2}$ and $y(1) = 1, y(1.1) = 0.996, y(1.2) = 0.986, y(1.3) = 0.972$.

5.4. Gaussian Quadrature

Evaluate the function at a set of optimally chosen points in the interval. We will choose $\{x_0, x_1, ...x_n\}$ belongs to [a,b] and coefficients c_i, So that the approximation

$$\int_a^b f(x)\,dx = \sum_{i=0}^{n} c_i f(x_i)$$

is exact for the largest class of polynomials possible.

We have already seen that the open Newton-Cotes formulas sometimes give us better 'bang-for-buck' than the closed formulas.

The mid-point formula uses only 1-point and is as accurate as the two-point trapezoidal rule.-Gaussian quadrature takes this one step further.

Example Suppose we want to find an optimal two-point formula $\int_{-1}^{1} f(x)\,dx = c_1 f(x_1) + c_2 f(x_2)$

Since we have 4 parameters to play with, we can generate a formula that is exact up to polynomials of degree 3. We get the following 4 equations.

306

$$\int_{-1}^{1} 1 dx = 2 = c_1 + c_2$$

$$\int_{-1}^{1} x dx = 0 = c_1 x_1 + c_2 x_2$$

$$\int_{-1}^{1} x^2 dx = \frac{2}{3} = c_1 x_1^2 + c_2 x_2^2$$

$$\int_{-1}^{1} x^3 dx = 0 = c_1 x_1^3 + c_2 x_2^3$$

$$c_1 = 1, c_2 = 1$$

$$x_1 = -\frac{\sqrt{3}}{3}, x_2 = \frac{\sqrt{3}}{3}$$

5.4.1. Higher order Gaussian Quadrature Formulas

We could obtain higher order formulas by adding more points, computing the integrals, and solving the resulting non-linear system of equations. But it gets very painful, very fast.

The Legendre Polynomials come to our rescue.

The Legendre polynomials $P_n(x)$ are orthogonal on [-1,1] with respect to the weight function $w(x) = 1, i.e.,$ $w(x) = 1, i.e.,$

$$\int_{-1}^{1} P_n(x) P_m(x) dx = \alpha_n \delta_{n,m} = \begin{cases} 0, m = n \\ \alpha_n, m = n \end{cases}$$

If $P(x)$ is a polynomial of degree less than n, then

$$\int_{-1}^{1} P_n(x) P(x) dx = 0$$

307

We will see Legendre polynomials in more detail later. For now, all we need to know is that they satisfy the property

$$\int_{-1}^{1} P_n(x) P_m(x) dx = \alpha_n \delta_{n,m}$$

And the first few Legendre polynomials are

$$P_0(x) = 1$$
$$P_1(x) = x$$
$$P_2(x) = x^2 - 1/3$$
$$P_3(x) = x^3 - 3x/5$$
$$P_4(x) = x^4 - 6x^2/7 + 3/35$$
$$P_5(x) = x^5 - 10x^3/9 + 5x/21$$

It turns out that the roots of the Legendre polynomials are the nodes in Gaussian quadrature.

Theorem Suppose that $\{x_1, x_2, \ldots x_n\}$ are the roots of the nth Legendre polynomial $P_n(x)$ and that for each $i = 1, 2 \ldots n$, the coefficients c_i are defined by

$$c_i = \int_{-1}^{1} \prod_{j=1}^{n} \frac{x - x_j}{x_i - x_j} dx$$

Theorem 5.1 If $P(x)$ is an polynomial of degree less than 2n, then

$$\int_{-1}^{1} P(x) dx = \sum_{i=1}^{n} c_i P(x_i)$$

Proof Let us first consider a polynomial $P(x)$ with degree less than n. $P(x)$ can be rewritten as an n-1 st Lagrange polynomial with nodes at the roots of the nth Legendre polynomial $P_n(x)$. This representation is exact since the error term involves the nth derivative of $P(x)$, which is zero. Hence

$$\int_{-1}^{1} P(x)\,dx = \int_{-1}^{1}\left[\sum_{i=1}^{n}\prod_{j=1}^{n}\frac{x-x_j}{x_i-x_j}P(x_j)\right]dx$$

$$= \sum_{i=1}^{n}\int_{-1}^{1}\left[\sum_{i=1}^{n}\prod_{j=1}^{n}\frac{x-x_j}{x_i-x_j}P(x_j)\,dx\right]P(x_i) = \sum_{i=1}^{n}c_i P(x_i)$$

Which verifies the result for polynomials of degree less than n.

www.ingramcontent.com/pod-product-compliance
Lightning Source LLC
Chambersburg PA
CBHW082104220526
45472CB00009B/2034